Studies in Systems, Decision and Control

Volume 353

The series "Studies in Systems, Decision and Control" (SSDC) covers both new developments and advances, as well as the state of the art, in the various areas of broadly perceived systems, decision making and control–quickly, up to date and with a high quality. The intent is to cover the theory, applications, and perspectives on the state of the art and future developments relevant to systems, decision making, control, complex processes and related areas, as embedded in the fields of engineering, computer science, physics, economics, social and life sciences, as well as the paradigms and methodologies behind them. The series contains monographs, textbooks, lecture notes and edited volumes in systems, decision making and control spanning the areas of Cyber-Physical Systems, Autonomous Systems, Sensor Networks, Control Systems, Energy Systems, Automotive Systems, Biological Systems, Vehicular Networking and Connected Vehicles, Aerospace Systems, Automation, Manufacturing, Smart Grids, Nonlinear Systems, Power Systems, Robotics, Social Systems, Economic Systems and other. Of particular value to both the contributors and the readership are the short publication timeframe and the world-wide distribution and exposure which enable both a wide and rapid dissemination of research output.

Indexed by SCOPUS, DBLP, WTI Frankfurt eG, zbMATH, SCImago.

All books published in the series are submitted for consideration in Web of Science.

More information about this series at http://www.springer.com/series/13304

John N. Mordeson · Sunil Mathew

Mathematics of Uncertainty for Coping with World Challenges

Climate Change, World Hunger, Modern Slavery, Coronavirus, Human Trafficking

Springer

John N. Mordeson
Department of Mathematics
Creighton University
Omaha, NE, USA

Sunil Mathew
Department of Mathematics
National Institute of Technology Calicut
Calicut, Kerala, India

ISSN 2198-4182 ISSN 2198-4190 (electronic)
Studies in Systems, Decision and Control
ISBN 978-3-030-68683-3 ISBN 978-3-030-68684-0 (eBook)
https://doi.org/10.1007/978-3-030-68684-0

This Springer imprint is published by the registered company Springer Nature Switzerland AG
The registered company address is: Gewerbestrasse 11, 6330 Cham, Switzerland

John Mordeson would like to dedicate the book to his wonderful grandchildren.

Sunil Mathew would like to dedicate the book to his soulmate Cini Mathew on her 43rd birthday.

Preface

It is stated in [51] that the coronavirus pandemic is a fire drill for our planet's future. Climate change will mean one emergency after another, year after year, as heat waves, floods, fires, and storms blow cascades of failures through our systems. Like this pandemic, climate change is also going to push on the networks that make up our civilization. Unlike the pandemic, its effects will be long lasting. The networks involved are the transportation network, energy network, economics network, and the health care network.

The US Centers for Disease Control and Prevention (CDC) estimates that three out of four new infectious diseases come from human–animal contact. The outbreaks of Ebola and other Coronaviruses such as MERS, for instance, were triggered by a jump from animal to human in disturbed natural habitats, [76]. These disruptions are caused in part by global warming.

Human livelihoods, stable economies, good health, and high quality of life all hinge on a stable climate and Earth system, and on a diversity of species and ecosystems. Yet biodiversity is declining faster than at any point in human history and time is running out to limit global temperature rise to well 2^0C and to aim for 1.5^0C. To do so, emissions must peak as soon as possible and decline sharply until 2050, but the world is not on that path, [13]. Climate change significantly impacts human trafficking and modern slavery. However, it is also the case that modern slavery has a negative impact on climate change. Accurate data concerning human trafficking and modern slavery is impossible to obtain. The purpose of this book is to analyze the crimes of human trafficking and modern slavery by the use of the mathematics of uncertainty.

In Chap. 1, we present the notation and basic material needed for the understanding of the book. This includes the notions of fuzzy subsets, fuzzy similarity measures, and fuzzy graphs. We also present the basics of evidence theory. This includes a discussion of the Analytic Hierarchy Process (AHP), Guiasu method, the Yen method, and Dempster-Shafer theory.

In Chap. 2, we rank countries with respect to their achievement of the Sustainable Development Goals (SDGs). We use a combination of the AHP, Guiasu, and Yen methods and another method called the Stakeholder method. We use two fuzzy similarity measures to determine a measure of how similar these rankings are and also a ranking of the countries given by Sustainable Development Report 2019, New York, Bertlesmann Stiftung and Sustainable development Solutions Network (SDSN), 2019.

In Chap. 3, we consider the risk of developed, emerging, and frontier market countries for their vulnerability to climate change. In [134], 67 countries were ranked according to their vulnerability. We place the findings in [134] into a mathematical framework using the mathematics of uncertainty. We use techniques to rank the countries as to their risk. We use two fuzzy similarity measures to determine the similarity of the ranking methods.

We turn our attention to the crimes of human trafficking and modern slavery in Chap. 4. We rank countries with respect to their achievement of the SDGs pertinent to these crimes. We also use two fuzzy similarity measures to determine the similarity of the ranking methods.

In Chap. 5, we consider the interconnections between climate change and modern slavery. It is known that climate change can lead to extreme events which cause an increase in immigration and thus an increase in modern slavery and human trafficking. However, modern slavery activities may further perpetuate climate change and contribute to significant Greenhouse Gas emissions. Our work in this chapter is based on [32]. We use the Guiasu method and the Dempster method to rank countries on their achievement pertinent to modern slavery.

Even though modern slavery is widespread, it is largely invisible. The Global Slavery Index 2018 is an attempt to measure the problem. In Chap. 6, we use the tools of mathematics of uncertainty to aid in analyzing the situation. We consider the findings of the Global Slavery Index concerning the prevalence of modern slavery in countries, the vulnerability of countries, and the government response of countries. We use fuzzy similarity measures to compare the rankings of the countries on The Global Slavery Index with respect to vulnerability, government response, and prevalence. We then develop several indices in fuzzy graph theory to measure the susceptibility of paths involving trafficking in Central America through Mexico to the United States.

In Chap. 7, we study the vulnerability of countries to climate change. We also examine the effects of climate change and the coronavirus on countries with respect to global hunger. We note that some models of the coronavirus can be placed in the setting of intuitionistic fuzzy mathematics.

In Chap. 8, we consider connectivity properties of fuzzy graphs. We introduce the notion of a container in a fuzzy graph. A container in a graph is a family of internally disjoint paths between a pair of vertices. A fuzzy container in a fuzzy graph is a family of internally disjoint strongest paths. We apply our results to examine connectivity properties with respect to human trafficking flow between regions in the world. We find the strength of connectedness between regions and fuzzy containers between them.

In Chap. 9, we continue our study of connectedness properties of fuzzy graphs. Generalizations for some existing vertex and edge connectivity in graphs are presented. New parameters are compared with older ones and generalized values are calculated for some of the major classes like cycles and trees in fuzzy graphs.

Contents

About the Authors

Dr. John N. Mordeson is a Professor Emeritus of Mathematics at Creighton University. He received his B.S., M.S., and Ph.D. from Iowa State University. He is a member of Phi Kappa Phi. He is the president of the Society for Mathematics of Uncertainty. He has published 18 books and 200 journal articles. He is on the editorial board of numerous journals. He has served as an external examiner of Ph.D. candidates from India, South Africa, Bulgaria, and Pakistan. He has refereed for numerous journals and granting agencies. He is particularly interested in applying the mathematics of uncertainty to combat the problem of human trafficking.

Dr. Sunil Mathew is a Faculty Member in the Department of Mathematics, NIT Calicut, India. He has acquired his masters from St. Joseph's College Devagiri, Calicut, and Ph.D. from the National Institute of Technology Calicut in the area of Fuzzy Graph Theory. He has published more than 100 research papers and written six books. He is a member of several academic bodies and associations. He is an editor and reviewer of several international journals. He has an experience of 20 years in teaching and research. His current research topics include fuzzy graph theory, bio-computational modeling, graph theory, fractal geometry, and chaos.

Chapter 1
Preliminaries

This chapter is concerned with the notation and concepts needed for the book. We assume the reader is familiar with basic set theory. We first consider notation. We let $\wedge$ denote minimum or infimum and $\vee$ denote maximum or supremum. Let $\mathbb{R}$ denote the set of real numbers and $\mathbb{N}$ denote the set of positive integers. We let $[0, 1]$ denote the closed unit interval. Let A be a subset of a set X. Then $X \backslash A$ (or A^c if the context is clear) denotes the complement of A in X. We let $|A|$ denote the cardinality of A and A^n denote the Cartesian cross product of A n times, where $n \in \mathbb{N}$. We let $\mathcal{P}(X)$ denote the power set of X.

1.1 Fuzzy Sets

Let X be a set and A a subset of X. The **characteristic function** of A is the function χ of X into $\{0, 1\}$ defined by $\chi(x) = 1$ if $x \in A$ and $\chi(x) = 0$ if $x \notin A$. The characteristic function can be used to indicate either members or nonmembers of A. This notion can be generalized in a way that introduces the notion of a fuzzy subset of X.

Lotfi Zadeh introduced the notion of a fuzzy subset of a set in [176]. Fuzzy subsets are sets with boundaries that are not precise. The membership in a fuzzy set is a matter of degree.

Definition 1.1.1 A fuzzy subset μ of X is a function of X into the closed interval $[0, 1]$.

Let μ be a fuzzy subset of a set X. For all $x \in X$, $\mu(x)$ can be thought of as the degree of membership of x in μ. We sometimes use the notation μ_A for a fuzzy subset of X, where A is thought of as a fuzzy set and μ_A gives the grade of membership of elements of X in A. At times, A may be merely a description of a fuzzy subset μ of X.

J. N. Mordeson and S. Mathew, *Mathematics of Uncertainty for Coping with World Challenges*, Studies in Systems, Decision and Control 353,
https://doi.org/10.1007/978-3-030-68684-0_1

We let $\mathcal{FP}(X)$ denote the **fuzzy power set** of X, i.e., the set of all fuzzy subsets of X.

Definition 1.1.2 Let μ be a fuzzy subset of X.

(1) Let $\alpha \in [0, 1]$. Define $\mu_a = \{x \in X | \mu(x) \geq \alpha\}$. We call μ_α and a-**cut** or a α-**level set.**

(2) The **support** of μ is defined to be the set $\text{Supp}(\mu) = \{x \in X | \mu(x) > 0\}$.

Definition 1.1.3 Let μ and ν be fuzzy subsets of X. Define μ^c, $\mu \cap \nu$, and $\mu \cup \nu$ as follows: $\forall x \in X$,

$$\mu^c(x) = 1 - \mu(x),$$
$$(\mu \cap \nu)(x) = \mu(x) \wedge \nu(x),$$
$$(\mu \cup \nu)(x) = \mu(x) \vee \nu(x).$$

Then μ^c is called the **(standard) complement** of μ, $\mu \cap \nu$ the **intersection** of μ and ν, and $\mu \cup \nu$ the **union** of μ and ν.

We can extend the notions of intersection and union to a family of fuzzy subsets of X. Let $\{\mu\}_{i \in I}$ be a family of fuzzy subsets of X, where I is an index set. Define $\cap_{i \in I}$ and $\cup_{i \in I}$ as follows: $\forall x \in X$,

$$(\cap_{i \in I} \mu_i)(x) = \wedge_{i \in I} \mu_i(x),$$
$$(\cup_{i \in I} \mu_i)(x) = \vee_{i \in I} \mu_i(x)$$

Thus if I is a finite set, say $I = \{1, 2, \ldots, n\}$, then $(\cap_{i \in I} \mu_i) = \mu_1 \cap \mu_2 \cap \cdots \cap \mu_n$ and $(\cup_{i \in I} \mu_i) = \mu_1 \cup \mu_2 \cup \cdots \cup \mu_n$. In this case, we sometimes write $(\cap_{i \in I} \mu_i)(x) = \mu_1(x) \wedge \mu_2(x) \wedge \cdots \wedge \mu_n(x)$ and $(\cup_{i \in I} \mu_i)(x) = \mu_1(x) \vee \mu_2(x) \vee \cdots \vee \mu_n(x)$.

The intersection of two fuzzy subsets of a set is specified in general by a binary operation on the unit interval; that is, a function of the form

$$i : [0, 1] \times [0, 1] \to [0, 1].$$

Definition 1.1.4 A fuzzy intersection (*t*-**norm**) is a binary relation i on the unit interval that satisfies the following properties: $\forall a, b, d \in [0, 1]$:
(1) $i(a, 1) = a$ (boundary condition);
(2) $b \leq d$ implies $i(a, b) \leq i(a, d)$ (monotonicity);
(3) $i(a, b) = i(b, a)$ (commutativity);
(4) $i(a, i(b, d)) = i(i(a, b), d)$ (associativity).

The following are examples of t-norms.

Example 1.1.5 Let $a, b \in [0, 1]$.

Standard intersection: $i(a, b) = a \wedge b$;
Algebraic product: $i(a, b) = ab$;
Bounded difference: $i(a, b) = 0 \vee (a + b - 1)$.

The general fuzzy union of two fuzzy subsets is specified by a function $u : [0, 1] \times [0, 1] \to [0, 1]$.

Definition 1.1.6 A fuzzy union (t-**conorm**) is a binary relation u on the unit interval that satisfies the following properties: $\forall a, b, d \in [0, 1]$:
(1) $u(a, 0) = a$ (boundary condition);
(2) $b \le d$ implies $u(a, b) \le u(a, d)$ (monotonicity);
(3) $u(a, b) = u(b, a)$ (commutativity);
(4) $u(a, u(b, d)) = u(u(a, b), d)$ (associativity).

The following are examples of t-conorms.

Example 1.1.7 Let $a, b \in [0, 1]$.
Standard union: $u(a, b) = a \vee b$;
Algebraic sum: $u(a, b) = a + b - ab$;
Bounded sum: $u(a, b) = 1 \wedge (a + b)$.

1.2 Evidence Theory

The following discussion is from [113]. One of the broadest frameworks for the representation of uncertainty is evidence theory. It is one of the oldest uncertainty frameworks. Its origins lie in the works of Dempster [40, 41] and Shafer [153] and are heavily influenced by probability theory. Evidence theory encompasses belief, plausibility, necessity, possibility, and probability and many other measures. Here we present Evidence Theory as it was originally characterized by Shafer.

Evidence theory is based on two fuzzy measures, belief measures and plausibility measures. Belief and plausibility measures can be conveniently characterized by a function m from the power set of a universal set X into the unit interval. We assume that X is finite in this section. The function $m : \mathcal{P}(X) \to [0, 1]$ is required to satisfy two conditions:
(1) $m(\emptyset) = 0$;
(2) $\sum_{A \in \mathcal{P}(X)} m(A) = 1$.

The function m is called a **basic probability assignment**. For each set $A \in \mathcal{P}(X)$, the value $m(A)$ expresses the proportion to which all available and relevant evidence supports the claim that a particular element of X belongs to the set A. This value $m(A)$ pertains solely to one set, set A; it does not imply any additional claims regarding subsets or supersets of A. It there is some additional subset of A, say $B \subseteq A$, it must be expressed by another value $m(B)$.

Given a basic probability assignment, m, every set $A \in \mathcal{P}(X)$ for which $m(A) \neq 0$ is called a **focal element**. The pair $(\mathcal{F}, m)$, where $\mathcal{F}$ denotes the set of all focal elements induced by m is called a **body of evidence** and we denote it by $\mathcal{B} = (\mathcal{F}, m)$.

From a basic probability assignment m, the corresponding belief measure and plausibility measure are determined for all sets $A \in \mathcal{P}(X)$ by the formulas

$$\mathrm{Bel}(A) = \sum_{B \subseteq A} m(B),$$

$$\mathrm{Pl}(A) = \sum_{B \cap A \neq \emptyset} m(B).$$

Thus the belief of a set A is the sum of all the evidence (basic probability) assigned to A or any subset of A. The plausibility of A is the sum of all the evidence (basic probability) that overlaps with A.

It can be shown that the plausibility of an event is one minus the belief of the compliment of that event, and vice versa. That is,

$$\mathrm{Bel}(A) = 1 - \mathrm{Pl}(A^c),$$

$$\mathrm{Pl}(A) = 1 - \mathrm{Bel}(A^c).$$

Since we can calculate the belief from the plausibility, and the plausibility from the belief, and both belief and plausibility can be derived from the basic probability assignment, we only need one formula to show that all three measures provide the same information.

Given a belief measure Bel, the corresponding basic probability assignment m is determined for all $A \in \mathcal{P}(X)$ by the formula

$$m(A) = \sum_{B \subset A} (-1)^{|A-B|} \mathrm{Bel}(B),$$

where $|A - B|$ is the cardinality of the set difference of A and B, as proven by Shafer [153]. Thus each of the three functions, m, Bel, and Pl is sufficient to determine the other two.

Total ignorance is expressed in evidence theory by $m(X) = 1$ and $m(A) = 0$ for all $A \subset X$. Full certainty is expressed by $m(\{x\}) = 1$ for one particular element x of X and $m(A) = 0$ for all $A \neq \{x\}$.

Guiasu Method

The Guiasu method is a process of reaching a verdict by probabilistic weighting the available evidence. The classical rules from decision theory proposed by Hooper, Dempster, Bayes, and Jeffrey are special cases of Guiasu's weighting process. The Guiasu method is a generalization of Dempster-Shafer theory [40, 41, 153] and makes use of fuzzy set theory.

A body of information induces a probability (credibility) distribution m on $\mathcal{P}(X)$, the set of all subsets of X. That is, m is a function of $\mathcal{P}(X)$ into the closed interval [0, 1], written $m : \mathcal{P}(X) \to [0, 1]$, such that $m(A) \geq 0 \forall A \in \mathcal{P}(X)$ and $\sum_{A \in \mathcal{P}(X)} m(A) = 1$. The class of focal subsets of X corresponding to m is denoted by $\mathcal{F}(X; m) = \{A | A \subseteq X, m(A) > 0\}$. A pair of dependent bodies of information, say i and j, induce a joint probability (credibility) distribution, namely $m_{ij} : \mathcal{P}(X) \times \mathcal{P}(X) \to [0, 1]$ such that $m_{ij}(A, B) \geq 0$ and $\sum_{A \subseteq X} m_{ij}(A, B) = 1$.

If the bodies of information are independent, then $m_{ij} = m_i m_j$. The corresponding class of focal pairs of subsets is $\mathcal{F}(X, X; m_{ij}) = \{(A, B)|A, B \subseteq X, m_{ij}(A, B) > 0\}$. The weights corresponding to the body of information for which m is the probability (credibility) distribution on $\mathcal{P}(X)$ are $w(\cdot|\cdot) : \mathcal{P}(X) \times \mathcal{F}(X; m) \rightarrow [0, \infty)$. The weighted body of information provides the new probability (credibility) distribution on $\mathcal{P}(X)$ given by $\mu(C) = \sum_{A \in \mathcal{F}(X;m)} w(C|A)m(A)$. We can generalize this procedure to formulate the weights $w_{ij}(\cdot|\cdot, \cdot)$ that are assigned to a mixed body of information inducing a joint probability (credibility) distribution induced on $\mathcal{P}(X)$ by the weighted (i, j)-th body of information, i.e.,

$$\mu_{ij}(C) = \sum_{(A,B) \in \mathcal{F}(X, X m_{ij})} w_{ij}(C|A, B)m_{ij}(A, B), C \in \mathcal{P}(X),$$

where $w_{ij}(C|A, B)$ is the weight of the subset C given $(A, B) \in \mathcal{F}(X, X; m_{ij})$. If the probability (credibility) distribution m on X is such that $\sum_{A \in \mathcal{F}(X;m)} m(A) = 1$ and $\forall A \in \mathcal{F}(X; m)$, $|A| = 1$, then it is called **probabilistic**.

The following discussion is explained via Sustainable Development Goals (SDGs). Given m subgoals (SDGs in this application) and n experts. Assume the experts assign numbers to each SDG a number with respect as their importance in the examination of the overarching goal (sustainability) to form a $m \times n$ matrix $W = [w_{ij}]$. When the columns of the matrix are normalized, we can consider that each column of the of the resulting matrix N to be a probability (credibility) distribution for each expert (t-norm in this application). These probability (credibility) distributions are probabilistic with the focal elements being singleton sets consisting of an SDG. The row averages provide the Guiasu weights, one for each SDG.

Theorem 1.2.1 ([129]) *The row averages of N give the Guiasu weights $w_i, i = 1, \ldots, m$.*

Analytic Hierarchy Process

The Analytical Hierarchy Process (AHP) is a multicriteria decision method introduced in [146, 147]. We consider a factor to be studied by the examination of subfactors of the factor. In our case, each expert E_j assigns a number, w_{ij}, to each subfactor (SDG), $G_i, i = 1, \ldots, m$, as to its importance with respect to the overarching goal (sustainability). The row average, w_i of each row of the matrix $W = [w_{ij}]$ is determined to form an $m \times n$-matrix R whose ij-th element is w_i/w_j. The columns of R are then normalized in order to form the $m \times n$-matrix N whose ij-th element is $(w_i/w_j)/\sum_{i=1}^{m} w_i/w_j = w_i/\sum_{i=1}^{m} w_i, i = 1, \ldots, m$. This row vector yields the weights for the subfactors (SDGs) for the linear equation of the overarching goal (sustainability), the dependent variable, in terms of the SDGs, the independent variables.

If the matrix W already has its columns normalized, then $w_i = \sum_{j=1}^{n} w_{ij}/n, i = 1, \ldots, m$. Since $\sum_{i=1}^{m} w_{ij} = 1, j = 1, \ldots, n$, it follows that $\sum_{i=1}^{m} w_i = 1$. Hence $w_i/\sum_{i=1}^{m} w_i = w_i$,i.e., w_i is the weight for the i-th SDG in the linear equation, $i = 1, \ldots, m$. It thus follows that if the columns of W are already normal, then the

Guiasu method (with probabilistic assignments) and the analytic hierarchy process yield the same weights. However, in general, the Guiasu weights and the analytic hierarchy process can have quite different weights [129].

Yen Method

Yen's method addresses the issue of managing imprecise and vague information in evidential reasoning by combining the Dempster-Shafer theory with fuzzy set theory [172]. Several researchers have extended the Dempster-Shafer theory to deal with vague information, but their extensions did not preserve an important principle that the belief and plausibility measures are lower and upper probabilities. Yen's method preserves this principle. Nevertheless, we use various measures of subsethood to determine belief functions. We do this to compare the results of the beliefs with Yen's method.

Yen's method is developed under the assumption that the focal elements are normal. If the fuzzy focal elements are not normal, he normalizes them.

Dempster Method

Evidence obtained in the same context from two independent sources and expressed by two basic probability assignments m_1 and m_2 over X can be combined in various ways into a joint basic probability assignment m_{12}. Here present the Dempster's rule of combination.

Suppose that m_1 is the basic probability assignment for a belief function Bel_1 over X and let $A_1, A_2, \ldots, A_k$ denote the focal elements of Bel_1. Let $m_2 be$ another basic probability assignment for a belief function Bel_2 over X with focal elements $B_1, B_2, \ldots, B_l$. Assume

$$\sum_{A_i \cap B_j = \emptyset} m_1(A_i) m_2(B_j) < 1. \tag{1.2.1}$$

The standard way of combining evidence is by the formula

$$m_{12}(C) = \frac{\sum_{A_i \cap B_j = C} m_1(A_i) m_2(B_j)}{1 - K}$$

for all nonempty subsets C of X and $m_{12}(\emptyset) = 0$, where

$$K = \sum_{A_i \cap B_j = \emptyset} m_1(A_i) m_2(B_j).$$

We next show that the basic probability assignment is normalized by the division by $1 - K$. We have that

$$\sum_{C\in\mathcal{P}(X)} m_{12}(C) = \sum_{C\in\mathcal{P}(X),C\neq\emptyset} m_{12}(C)\frac{\sum_{A_i\cap B_j=C} m_1(A_i)m_2(B_j)}{1-K}$$
$$\frac{\sum_{C\in\mathcal{P}(X),C\neq\emptyset} m_{12}(C)\sum_{A_i\cap B_j=C} m_1(A_i)m_2(B_j)}{1-K}$$
$$= 1$$

since

$$1-K = 1-\sum_{A_i\cap B_j=\emptyset} m_1(A_i)m_2(B_j)$$
$$= \sum_{C\in\mathcal{P}(X),C\neq\emptyset} m_{12}(C)\sum_{A_i\cap B_j=C} m_1(A_i)m_2(B_j).$$

The core of the belief function given by m_{12} is equal to the intersection of the cores of Bel_1 and Bel_2. The belief function associated with m_{12} is called the **orthogonal sum** or the **joint belief** of Bel_1 and Bel_2 and is denoted by $\mathrm{Bel}_1\oplus\mathrm{Bel}_2$. If Eq. (1.2.1) does not hold, we say that the orthogonal sum $\mathrm{Bel}_1\oplus\mathrm{Bel}_2$ does not exist.

By repeatedly applying Dempster's rule of combination, we can combine any number of belief functions.

1.3 Stakeholder Method

In [133], Stakeholder Forum created a methodology to enable scores to be assigned to each of the different targets and goals according to their significance in different contexts. The method uses a number of assessors to assign their own independent scores of the significance of each of the proposed targets in the implementation context in question, according to three separate criteria. The three criteria proposed are applicability, implementability, and their transformational impact (both in the country concerned and for the world as a whole), [133]. The assessors' scores are then aggregated and averaged to give an overall score for each target, and then combined to give an average score for each goal.

The general effect is to give the highest scores to those targets and goals which are both clearly applicable and implementable in the country in question and which represent the biggest transformational challenge. Conversely, lower scores are given to targets and goals which are less applicable or implementable in a particular country for reasons given in [133].

The table in [[133], Annex 2, p. 12] present the individual category scores and the overall scores for each goal and target. These were obtained by averaging the collective scores from the assessors. The scores given were out of a maximum of 2 for individual category scores and a maximum of 8 for overall scores. An overall score was then obtained for each target by multiplying the scores given to each of the three categories. Multiplication was used to emphasize that for a goal or target to score highly, it must meet all three criteria. We divided the individual category scores by 2 so that the resulting scores were out of a maximum of 1. The overall scores were divided by 8 so the resulting scores were out of a maximum of 1. In this way, we have placed the scoring in a fuzzy logic setting. We note that multiplication is a particular t-norm in fuzzy logic.

1.4 Fuzzy Similarity Measures

In this section, we take a brief look at fuzzy similarity measures. We are interested in measuring the similarity between rankings.

Definition 1.4.1 Let S be a function of $\mathcal{FP}(X) \times \mathcal{FP}(X)$ into $[0, 1]$. Then S is called a **fuzzy similarity measure** on $\mathcal{FP}(X)$ if the following properties hold: $\forall \mu, \nu, \rho \in \mathcal{FP}(X)$,
(1) $S(\mu, \nu) = S(\nu, \mu)$;
(2) $S(\mu, \nu) = 1$ if and only if $\mu = \nu$;
(3) If $\mu \subseteq \nu \subseteq \rho$, then $S(\mu, \rho) \leq S(\mu, \nu) \wedge S(\nu, \rho)$;
(4) If $S(\mu, \nu) = 0$, then $\forall x \in X$, $\mu(x) \wedge \nu(x) = 0$.

We apply fuzzy similarity measures to rankings of members of a finite set. Suppose that X is a finite set with n elements. Let A be a one-to-one function of X into $\{1, 2, \ldots, n\}$. Then A is called a **ranking** of X. Define the fuzzy subset μ_A of X as follows: $\forall x \in X$, $\mu_A(x) = A(x)/n$. We wish to consider the similarity of two rankings of X by the use of similarity measures.

Much of our discussion is from [113, 166].

Example 1.4.2 Let μ_A and μ_B be the fuzzy subsets of X associated with two rankings A and B of X, respectively. Then M, L, and S are fuzzy similarity measures.

$$M(\mu_A, \mu_B) = \frac{\sum_{x \in X} \mu_A(x) \wedge \mu_B(x)}{\sum_{x \in X} \mu_A(x) \vee \mu_B(x)};$$

$$L(\mu_A, \mu_B) = 1 - \vee_{x \in X} |\mu_A(x) - \mu_B(x)|;$$

$$S(\mu_A, \mu_B) = 1 - \frac{\sum_{x \in X} |\mu_A(x) - \mu_B(x)|}{\sum_{x \in X} (\mu_A(x) + \mu_B(x))}.$$

Lemma 1.4.3 *Let $a, b, c \in [0, 1]$. Then $(a \vee c) \wedge (b \vee c) = (a \wedge b) \vee c$.*

Lemma 1.4.4 *Let $x, y, z \in (0, 1]$ be such that $x \leq y$. Then $(x \vee z)/(y \vee z) \geq x/y$.*

Proof The result follows by considering the three cases $z \leq x \leq y$, $x \leq z \leq y$, and $x \leq y \leq z$. □

Theorem 1.4.5 *Let M be the fuzzy similarity measure of Example 1.4.2. Then $M(\mu_A \cup \mu_C, \mu_B \cup \mu_C) \geq M(\mu_A, \mu_B)$.*

Proof Applying the previous lemmas, we have that

$$\begin{aligned} M(\mu_A \cup \mu_C, \mu_B \cup \mu_C) &= \frac{\sum_{x\in X}(\mu_A(x) \vee \mu_C(x)) \wedge (\mu(_B(x) \vee \mu_C(x))}{\sum_{x\in X} \mu_A(x) \vee \mu_C(x) \vee \mu_B(x) \vee \mu_C(x)} \\ &= \frac{\sum_{x\in X}(\mu_A(x) \wedge \mu_B(x)) \vee \mu_C(x))}{\sum_{x\in X} \mu_A(x) \vee \mu_B(x) \vee \mu_C(x)} \\ &\geq \frac{\sum_{x\in X} \mu_A(x) \wedge \mu_B(x)}{\sum_{x\in X} \mu_A(x) \vee \mu_B(x)} \\ &= M(\mu_A, \mu_B). \end{aligned}$$

□

Lemma 1.4.6 *Let* $x, y, z \in [0, 1]$. *Then* $|(x \vee z) - (y \vee z)|/(x \vee z) + (y \vee z) \leq |x - y|/(x + y)$.

Proof There is no loss in generality in assuming $x \geq y$. Suppose $z \geq x \geq y$. Then $(x \vee z) - (y \vee z) = z - z \leq x - y$. Assume $x \geq y \geq z$. Then $(x \vee z) - (y \vee z) = x - y$. Suppose $x \geq z \geq y$. Then

$$\begin{aligned} y &\leq z \\ 2xy &\leq 2xz \\ xy - zx &\leq -xy + zx \\ x^2 + xy - zx - zy &\leq x^2 - xy + zx - zy \\ (x - z)(x + y) &\leq (x + z)(x - y) \\ \frac{x - z}{x + z} &\leq \frac{x - y}{x + y} \\ \frac{x \vee z - y \vee z}{x \vee z + y \vee z} &\leq \frac{x - y}{x + y}. \end{aligned}$$

□

Theorem 1.4.7 *Let S be the fuzzy similarity measure of Example 1.4.2. Then* $S(\mu_A \cup \mu_C, \mu_B \cup \mu_C) \geq S(\mu_A, \mu_B)$.

Proof We have by the Lemma 1.4.6 that

$$\begin{aligned} S(\mu_A \cup \mu_C, \mu_B \cup \mu_C) &= 1 - \frac{\sum_{x\in X} |(\mu_A(x) \vee \mu_C(x)) - (\mu(_B(x) \vee \mu_C(x))|}{\sum_{x\in X} \mu_A(x) \vee \mu_C(x) + \mu_B(x) \vee \mu_C(x)} \\ &\geq 1 - \frac{\sum_{x\in X} |\mu_A(x) - \mu(_B(x)|}{\sum_{x\in X} \mu_A(x) + \mu_B(x)} \\ &= S(\mu_A, \mu_B). \end{aligned}$$

□

Lemma 1.4.8 *Let $a, b, c \in [0, 1]$ be such that $a \geq b$. Then*
(1) $a - b \geq a \vee c - b \vee c$;
(2) $a - b \geq a \wedge c - b \wedge c$.

Proof (1) Suppose (i) $c \geq a \geq b$. Then $a \vee c - b \vee c = c - c \leq a - b$. Suppose (ii) $a \geq c \geq b$. Then $a \vee c - b \vee c = a - c \leq a - b$. Suppose (iii) $a \geq b \geq c$. Then $a \vee c - b \vee c = a - b$.

(2) Suppose (i) $c \geq a \geq b$. Then $a \wedge c - b \wedge c = a - b$. Suppose (ii) $a \geq c \geq b$. Then $a \wedge c - b \wedge c = c - b \leq a - b$. Suppose (iii) $a \geq b \geq c$. Then $a \vee c - b \vee c = c - c = 0 \leq a - b$. □

Theorem 1.4.9 *Let μ_A, μ_B, μ_C be fuzzy subsets of X. Let L be the fuzzy similarity measure of Example 1.4.2. Then*
(1) $L(\mu_A, \mu_B) \leq L(\mu_A \cup \mu_C, \mu_B \cup \mu_C)$;
(2) $L(\mu_A, \mu_B) \leq L(\mu_A \cap \mu_C, \mu_B \cap \mu_C)$.

Proof (1) By (1) of Lemma 1.4.8, we have for all $x \in X$ that

$$|\mu_A(x) - \mu_B(x)| \geq |\mu_A(x) \vee \mu_C(x) - \mu_B(x) \vee \mu_C(x)|$$
$$1 - |\mu_A(x) - \mu_B(x)| \leq 1 - |\mu_A(x) \vee \mu_C(x) - \mu_B(x) \vee \mu_C(x)|.$$

A similar argument holds for (2). □

We determine some useful properties of similarity measures.

Let $A = \{a_1, \ldots, a_n\}$ be a set of real numbers such that $a_1 < \cdots < a_n$. Let π and ρ be permutations of A. We consider the following similarity relation,

$$\frac{\sum_{i=1}^{n} \pi(a_i) \wedge \rho(a_i)}{\sum_{i=1}^{n} \pi(a_i) \vee \rho(a_i)}.$$

Consider $\sum_{i=1}^{n} a_i \wedge a_{n-i+1}$ and $\sum_{i=1}^{n} a_i \vee a_{n-i+1}$. Let n be even. Then $a_i \wedge a_{n-i+1} = a_i$ and $a_i \vee a_{n-i+1} = a_{n-i+1}$ for $i = 1, \ldots, n/2$. Thus

$$\frac{\sum_{i=1}^{n} a_i \wedge a_{n-i+1}}{\sum_{i=1}^{n} a_i \vee a_{n-i+1}} = \frac{2\sum_{i=1}^{n/2} a_i}{2\sum_{i=1}^{n/2} a_{n-i+1}}. \tag{1.4.1}$$

That is, the numerator is twice the sum of the smallest elements while the denominator is twice the sum of the largest elements. Consequently,

$$\frac{\sum_{i=1}^{n} a_i \wedge a_{n-i+1}}{\sum_{i=1}^{n} a_i \vee a_{n-i+1}} < \frac{\sum_{i=1}^{n} \pi(a_i) \wedge \rho(a_i)}{\sum_{i=1}^{n} \pi(a_i) \vee \rho(a_i)}. \tag{1.4.2}$$

Suppose n is odd. Then a similar argument shows (1.4.2) holds since there will be a middle element which will appear twice, once in the numerator and once on the denominator of

$$\frac{\sum_{i=1}^{n} a_i \wedge a_{n-i+1}}{\sum_{i=1}^{n} a_i \vee a_{n-i+1}}.$$

Suppose we consider n countries and they have been ranked twice 1 through n with no ties. We wish to consider the similarity of their rankings using the above fuzzy similarity operation. We can accomplish this by mapping the countries to their rank divided by n. For example, let $\mathcal{C}$ denote a set of n countries and if country C is ranked i, then we define the fuzzy subset μ of $\mathcal{C}$ by $\mu(C) = i/n$. Let μ and ν be two such fuzzy subsets of $\mathcal{C}$. Then

$$M(\mu,\nu) = \frac{\sum_{i=1}^{n} \mu(C_i) \wedge \nu(C_i)}{\sum_{i=1}^{n} \mu(C_i) \vee \nu(C_i)} = \frac{\sum_{i=1}^{n} n\mu(C_i) \wedge n\nu(C_i)}{\sum_{i=1}^{n} n\mu(C_i) \vee n\nu(C_i)}.$$

Consequently, there is no loss in generality in assuming that we are measuring the similarity of two rankings using the integers $1,\ldots,n$. Hence in the following, we let $a_i = i, i = 1,\ldots,n$. Then $\sum_{i=1}^{n} \pi(a_i) \wedge \rho(a_i) + \sum_{i=1}^{n} \pi(a_i) \vee \rho(a_i) = (n+1)n$ for any two permutations π and ρ.

We now consider $\frac{\sum_{i=1}^{n} a_i \wedge a_{n-i+1}}{\sum_{i=1}^{n} a_i \vee a_{n-i+1}}$, where the a_i are the integers 1 through n. Then by (1.4.1) (with n even)

$$\begin{aligned}
\frac{\sum_{i=1}^{n} a_i \wedge a_{n-i+1}}{\sum_{i=1}^{n} a_i \vee a_{n-i+1}} &= \frac{\sum_{i=1}^{n/2} a_i}{\sum_{i=1}^{n/2} a_{n-i+1}} \\
&= \frac{(n/2+1)(n/2)}{(n+1)n - (n/2+1)(n/2)} \\
&= \frac{1}{\frac{(n+1)n}{(n/2+1)(n/2)} - 1} \\
&= \frac{1}{4(n+1)/(n+2) - 1}.
\end{aligned}$$

Now $\lim_{n\to\infty} \frac{1}{4(n+1)/(n+2)-1} = 1/3$ (letting the terms in the limit equal $1/3$ if n is odd). A similar argument shows that if n is odd, $\lim_{n\to\infty} \frac{\sum_{i=1}^{n} a_i \wedge a_{n-i+1}}{\sum_{i=1}^{n} a_i \vee a_{n-i+1}} = 1/3$ (letting the terms in the limit be $1/3$ if n is even). Hence, we have in general that

$$\lim_{n\to\infty} \frac{\sum_{i=1}^{n} a_i \wedge a_{n-i+1}}{\sum_{i=1}^{n} a_i \vee a_{n-i+1}} = 1/3. \tag{1.4.3}$$

Therefore, we have by (1.4.2) and (1.4.3), a bench mark to determine how similar two rankings are.

We next show that $\frac{\sum_{i=1}^{n} a_i \wedge a_{n-i+1}}{\sum_{i=1}^{n} a_i \vee a_{n-i+1}}$ is a decreasing sequence whose elements are greater than $\frac{1}{3}$, but converge to $\frac{1}{3}$.

Let n be even. Then

$$\frac{\sum_{i=1}^{n} a_i \wedge a_{n-i+1}}{\sum_{i=1}^{n} a_i \vee a_{n-i+1}} = \left(\frac{n}{2}+1\right)\left(\frac{n}{2}\right) / \left[(n+1)n - \left(\frac{n}{2}+1\right)\left(\frac{n}{2}\right)\right]$$
$$= \left(\frac{n}{2}+1\right)\left(\frac{n}{2}\right) / n\left[n+1-\frac{n+2}{4}\right]$$
$$= \frac{n+2}{4} / \frac{3n+2}{4}$$
$$= \frac{n+2}{3n+2} \qquad (1.4.4)$$

Let n be odd. Let $N = (\frac{n-1}{2}+1)(\frac{n-1}{2}) + \frac{n+1}{2}$. Then

$$\frac{\sum_{i=1}^{n} a_i \wedge a_{n-i+1}}{\sum_{i=1}^{n} a_i \vee a_{n-i+1}} = \left[\left(\frac{n-1}{2}+1\right)\left(\frac{n-1}{2}\right)+\frac{n+1}{2}\right] / \left[2\left(\frac{n+1}{2}\right)(n) - N\right]$$
$$= \frac{\frac{n+1}{2}\frac{n-1}{2} + \frac{n+1}{2}}{(n+1)(n) - N}$$
$$= \frac{\frac{n+1}{2}\frac{n+1}{2}}{(n+1)n - \frac{n+1}{2}\frac{n+1}{2}}$$
$$= \frac{\frac{n+1}{2}\frac{n+1}{2}}{(n+1)(n - \frac{n+1}{4})}$$
$$= \frac{n+1}{3n-1}. \qquad (1.4.5)$$

Let $n = 2k$. Then in (1.4.4), $\frac{n+2}{3n+2} = \frac{2k+2}{6k+2} = \frac{k+1}{3k+1}$. Let $n = 2k+1$. Then in (1.4.5), $\frac{n+1}{3n-1} = \frac{2k+2}{6k+3-1} = \frac{k+1}{3k+1}$. That is, $\frac{\sum_{i=1}^{n} a_i \wedge a_{n-i+1}}{\sum_{i=1}^{n} a_i \vee a_{n-i+1}}$ is the same for n even and $n+1$ odd. However note that for n odd, $\frac{n+1}{3n-1} > \frac{n+1+2}{3(n+1)+2} = \frac{n+3}{3n+5}$. That is, for n odd, $\frac{\sum_{i=1}^{n} a_i \wedge a_{n-i+1}}{\sum_{i=1}^{n} a_i \vee a_{n-i+1}}$ is strictly greater than $\frac{\sum_{i=1}^{n} a_i \wedge a_{n-i+1}}{\sum_{i=1}^{n} a_i \vee a_{n-i+1}}$ for the next even. Clearly, $\frac{\sum_{i=1}^{n} a_i \wedge a_{n-i+1}}{\sum_{i=1}^{n} a_i \vee a_{n-i+1}} > \frac{1}{3}$ for all n. Thus $\frac{\sum_{i=1}^{n} a_i \wedge a_{n-i+1}}{\sum_{i=1}^{n} a_i \vee a_{n-i+1}}$ is a decreasing sequence whose elements are greater than $\frac{1}{3}$, but converges to $\frac{1}{3}$.

We next consider the fuzzy similarity measure

$$S(\mu, \nu) = 1 - \frac{\sum_{i=1}^{n} |\mu(x_i) - \nu(x_i)|}{\sum_{i=1}^{n}(\mu(x_i) + \nu(x_i))}$$

Assume $\{a_1, a_2, \ldots, a_n\} = \{1, 2, \ldots, n\} = \{b_1, b_2, \ldots, b_n\}$.
Consider

$$S = 1 - \frac{\sum_{i=1}^{n} |a_i - b_i|}{\sum_{i=1}^{n}(a_i + b_i)}.$$

We are first interested in the smallest value S can take on.

First suppose that n is even. Then $\sum_{i=1}^{n}(a_i + b_i) = \sum_{i=1}^{n} a_i + \sum_{i=1}^{n} b_i = n(n+1)/2 + n(n+1)/2 = n(n+1)$. The largest value $\sum_{i=1}^{n} |a_i - b_i|$ can be is when $\forall i \in \{1, 2, \ldots, n\}$, either $a_i \in \{1, \ldots, \frac{n}{2}\}$ and $b_i \in \{\frac{n}{2}+1, \ldots, n\}$ or $b_i \in \{1, \ldots, \frac{n}{2}\}$ and $a_i \in \{\frac{n}{2}+1, \ldots, n\}$. This holds since the a_i are distinct and vary over all $\{1, \ldots, n\}$ and similarly for the b_i. (It doesn't matter how the a_i and the b_i are paired. That is, $\sum_{i=1}^{n} |a_i - b_i| = 2(\frac{n}{2} + 1 + \frac{n}{2} + 2 \ldots + n - (1 + 2 + \cdots + \frac{n}{2}))$.) We consider $a_i = n - i + 1, \ldots, 2, 1$ and $b_i = i, i = 1, 2, \ldots, n$. Then

$$\begin{aligned}
\sum_{i=1}^{n} |a_i - b_i| &= 2(n - 1 + n - 3 + \cdots + 1) \\
&= 2\left(\frac{n(n-1)}{2} - (n - 2 + n - 4 + \cdots + 2)\right) \\
&= 2\left(\frac{n(n-1)}{2} - 2\left(\frac{n-2}{2} + \frac{n-4}{2} + \cdots + 1\right)\right) \\
&= 2\left(\frac{n(n-1)}{2} - 2\left(\frac{n-2}{2} + 1\right)\left(\frac{n-2}{2}\right)/2\right) \\
&= n(n-1) - n\frac{n-2}{2} \\
&= n^2 - n - \frac{n^2}{2} + n \\
&= \frac{n^2}{2}.
\end{aligned}$$

Thus

$$\begin{aligned}
S &= 1 - \frac{n^2/2}{n(n+1)} \\
&= \frac{(n^2/2) + n}{n(n+1)} \\
&= \frac{n/2 + 1}{n+1}.
\end{aligned} \tag{1.4.6}$$

Hence $\lim_{n \to \infty} S = \frac{1}{2}$.

Suppose n is odd. The a similar argument shows $\lim_{n \to \infty} S = \frac{1}{2}$. Since follows since the middle terms of the a_i and the b_i are equal and subtract to 0.

We supply the details. As before, $\sum_{i=1}^{n}(a_i + b_i) = \sum_{i=1}^{n} a_i + \sum_{i=1}^{n} b_i = n(n+1)/2 + n(n+1)/2 = n(n+1)$. The largest value $\sum_{i=1}^{n} |a_i - b_i|$ can be is when $\forall i \in \{1, 2, \ldots, n\}$, either $a_i \in \{1, \ldots, \frac{n+1}{2}\}$ and $b_i \in \{\frac{n+1}{2} + \frac{n+1}{2} + 1, \ldots, n\}$ or $b_i \in \{1, \ldots, \frac{n+1}{2}\}$ and $a_i \in \{\frac{n+1}{2}, \frac{n+1}{2} + 1, \ldots, n\}$. This holds since the a_i are distinct and vary over all $\{1, \ldots, n\}$ and similarly for the b_i. It doesn't matter how the a_i and the b_i are paired. That is,

$$\begin{aligned}
&\sum_{i=1}^{n} |a_i - b_i| \\
&= 2\left(\frac{n+1}{2} + \frac{n+1}{2} + 1 + \frac{n+1}{2} + 2 + \cdots + \frac{n+1}{2} + \frac{n-1}{2} - \left(1 + 2 + \cdots + \frac{n+1}{2}\right)\right) \\
&= 2\left(\frac{n+1}{2} + \cdots + \frac{n+1}{2}\right) \\
&= 2\left(\frac{n+1}{2}\right)\left(\frac{n-1}{2}\right) \\
&= \frac{(n^2-1)}{2}.
\end{aligned}$$

Thus

$$\begin{aligned}
S &= 1 - \frac{\frac{n^2-1}{2}}{n(n+1)} \\
&= 1 - \frac{n-1}{2n} \\
&= \frac{1}{2} + \frac{1}{2n}.
\end{aligned} \tag{1.4.7}$$

Equations (1.4.4)–(1.4.7) give the smallest value $M(\mu, \nu)$ and $S(\mu, \nu)$ can take on for a given n. In general, $M(\mu, \nu)$ and $S(\mu, \nu)$ are bounded below by $\frac{1}{3}$ and $\frac{1}{2}$, respectively. If we wish to calculate a value for $M(\mu, \nu)$ and $S(\mu, \nu)$ in which the values are bounded below by 0, we can use the formulas

$$\frac{M(\mu, \nu) - \frac{n+2}{3n+2}}{1 - \frac{n+2}{3n+2}} \text{ and } \frac{S(\mu, \nu) - \frac{n/2+1}{n+1}}{1 - \frac{n/2+1}{n+1}}$$

if n is even and

$$\frac{M(\mu, \nu) - \frac{n+1}{3n-1}}{1 - \frac{n+1}{3n-1}} \text{ and } \frac{S(\mu, \nu) - (\frac{1}{2} + \frac{1}{2n})}{1 - (\frac{1}{2} + \frac{1}{2n})}$$

if n is odd.

We next consider the similarity measure

$$S_1(\mu, \nu) = \frac{1}{n} \sum_{i=1}^{n} \frac{\mu(C_i) \wedge \nu(C_i)}{\mu(C_i) \vee \nu(C_i)}.$$

Example 1.4.10 Let $a_1 = 4, a_2 = 3, a_3 = 2, a_4 = 4$ and $b_1 = 1, b_2 = 2, b_3 = 3, b_4 = 4$. Then

$$\begin{aligned} S_1 &= \frac{1}{4}\sum_{i=1}^{4}\frac{a_i \wedge b_i}{a_i \vee b_i} \\ &= \frac{1}{4}\left(\frac{1}{4}+\frac{2}{3}+\frac{2}{3}+\frac{1}{4}\right) \\ &= \frac{1}{4}\left(\frac{11}{6}\right) \\ &= \frac{11}{24} = \frac{22}{48}. \end{aligned}$$

Let $c_1 = 2, c_2 = 1, c_3 = 3, c_4 = 4$. Then

$$\begin{aligned} S_1 &= \frac{1}{4}\sum_{i=1}^{4}\frac{a_i \wedge c_i}{a_i \vee c_i} \\ &= \frac{1}{4}\left(\frac{2}{4}+\frac{1}{3}+\frac{2}{3}+\frac{1}{4}\right) \\ &= \frac{1}{4}\left(\frac{7}{4}\right) \\ &= \frac{7}{16} = \frac{21}{48}. \end{aligned}$$

Thus it is not the case $a_1 = 4, a_2 = 3, a_3 = 2, a_4 = 4$ and $b_1 = 1, b_2 = 2, b_3 = 3, b_4 = 4$ yields the smallest similarity value.

The common existing properties of fuzzy similarity measures are:
(1) $S(A, B) = S(B, A) \forall A, B \in FP(X)$;
(2) $0 \leq S(A, B) \leq 1 \forall A, B \in FP(X)$;
(3) $\forall A, B \in FP(X)$, $S(A, B) = 1$ if and only if $A = B$;
(4) $\forall A, B, C \in FP(X)$, if $A \subseteq B \subseteq C$, then $S(A, B) \geq S(A, C)$ and $S(B, C) \geq S(A, C)$.

The common properties of distance measures are:
(1) $d(A, B) = d(B, A) \forall A, B \in FP(X)$;
(2) $0 \leq d(A, B) \leq 1 \forall A, B \in FP(X)$;
(3) $\forall A, B \in FP(X), d(A, B) = 0$ if and only if $A = B$;
(4) $\forall A, B, C \in FP(X)$, if $A \subseteq B \subseteq C$, then $d(A, B) \leq d(A, C)$ and $d(B, C) \leq d(A, C)$.

Let S and d be functions of $FP(X) \times FP(X)$ into $[0, 1]$. If S and d satisfy the above properties, respectively, then we call S a fuzzy similarity measure and d a fuzzy distance measure.

Let c be an involutive fuzzy complement.

Theorem 1.4.11 (1) *If S is a fuzzy similarity measure, then S^c is a fuzzy distance measure.*

(2) *If d is a fuzzy distance measure, then d^c is a fuzzy similarity measure.*

We see that the above theorem allows us to consider intuitionistic or neutrosophic fuzzy sets.

Theorem 1.4.12 $S(A, B) \geq M(A, B)$ *with* $S(A, B) = M(A, B)$ *if and only if* $a_i = b_i, i = 1, \ldots, n$.

Proof We have that

$$\begin{aligned} S(A, B) &= 1 - \frac{\sum_{i=1}^{n} |a_i - b_i|}{\sum_{i=1}^{n} (a_i + b_i)} = \frac{\sum_{i=1}^{n} (a_i + b_i) - \sum_{i=1}^{n} |a_i - b_i|}{\sum_{i=1}^{n} (a_i + b_i)} \\ &= \frac{\sum_{i=1}^{n} (a_i \vee b_i + a_i \wedge b_i) - \sum_{i=1}^{n} (a_i \vee b_i - a_i \wedge b_i)}{\sum_{i=1}^{n} (a_i \vee b_i + a_i \wedge b_i)} \\ &= \frac{\sum_{i=1}^{n} (a_i \wedge b_i) + \sum_{i=1}^{n} (a_i \wedge b_i)}{\sum_{i=1}^{n} (a_i \vee b_i) + \sum_{i=1}^{n} (a_i \wedge b_i)} \\ &\geq \frac{\sum_{i=1}^{n} (a_i \wedge b_i)}{\sum_{i=1}^{n} (a_i \vee b_i)} = M(A, B). \end{aligned}$$

□

1.5 Implication Operations and Similarity Operations

In this section, we define and examine similarity measures in terms of implication operators. Our discussion is from [16, 166].

Definition 1.5.1 Let $\mathcal{I}$ be a function of $[0, 1] \times [0, 1]$ into $[0, 1]$ such that $\mathcal{I}(0, 0) = \mathcal{I}(0, 1) = \mathcal{I}(1, 1) = 1$ and $\mathcal{I}(1, 0) = 0$. Then $\mathcal{I}$ is called an **implication operation**.

Let μ_A and μ_B be two fuzzy subsets of a set X. Let $\mathcal{I}$ be an implication operator. Then the degree to which μ_A is a subset of μ_B is defined to be

$$\wedge\{\mathcal{I}(\mu_A(x), \mu_B(x)) | x \in X\}.$$

Define the fuzzy subset $\subseteq_I$of $\mathcal{FP}(X) \times \mathcal{FP}(X)$ by $\forall \mu_A, \mu_B \in \mathcal{FP}(X), \subseteq_I (\mu_A, \mu_B) = \wedge\{\mathcal{I}(\mu_A(x), \mu_B(x)) | x \in X\}$.

Definition 1.5.2 Let $\mathcal{I}$ be an implication operator. Define the fuzzy subset $E_{\mathcal{I}}$ of $\mathcal{FP}(X) \times \mathcal{FP}(X)$ by for all $\mu_A, \mu_B \in \mathcal{FP}(X)$,

$$E_{\mathcal{I}}(\mu_A, \mu_B) = \wedge\{\wedge\{\mathcal{I}(\mu_A(x), \mu_B(x)) | x \in X\}, \wedge\{\mathcal{I}(\mu_B(x), \mu_A(x)) | x \in X\}\}.$$

Then is called the **degree of sameness** of μ_A and μ_B,[16].

Let $\mathcal{T}$ denote a t-norm. Then there exists an implication operator $\mathcal{I}_{\mathcal{T}}$ defined by $\mathcal{I}_{\mathcal{T}}(x, y) = \vee\{z|z \in [0, 1] \text{ and } \mathcal{T}(x, z) \leq y\}$. The following implication operators can be determined by a suitable t-norm, [166].

Example 1.5.3 Let $x, y \in [0, 1]$.

(1) Godel implication operator: $\mathcal{I}(x, y) = \begin{cases} 1 \text{ if } x \leq y, \\ y \text{ otherwise.} \end{cases}$

(2) Goguen implication operator: $\mathcal{I}(x, y) = \begin{cases} 1 \text{ if } x \leq y, \\ y/x \text{ otherwise.} \end{cases}$

(3) Luckasiewicz implication operator: $\mathcal{I}(x, y) = \wedge\{1 - x + y, 1\}$.

Definition 1.5.4 Let $\mathcal{I}$ be an implication operator. Then $\mathcal{I}$ is called **hybrid monotonous** if $\mathcal{I}(x, \ )$ is nondecreasing for all $x \in [0, 1]$ and $\mathcal{I}(\ , y)$ is nonincreasing for all $y \in [0, 1]$.

The implication operators in the previous example are hybrid monotonous.

Proposition 1.5.5 ([166]) *Let* A *be a finite subset of* $[0, 1]$ *and* $b \in [0, 1]$. *Let* $\mathcal{I}$ *be a hybrid monotonous implication operator. Then*

(1) $\mathcal{I}(\vee\{a|a \in A\}, b) = \wedge\{\mathcal{I}(a, b)|a \in A\}$;
(2) $\mathcal{I}(\wedge\{a|a \in A\}, b) = \vee\{\mathcal{I}(a, b)|a \in A\}$;
(3) $\mathcal{I}(b, \vee\{a|a \in A\}) = \vee\{\mathcal{I}(a, b)|a \in A\}$;
(4) $\mathcal{I}(b, \wedge\{a|a \in A\}) = \wedge\{\mathcal{I}(b, a)|a \in A\}$.

Lemma 1.5.6 ([166]) *Let* $\mathcal{I}$ *be the Luckasiewicz implication operator. Let* $a, b \in [0, 1]$. *Then*

$$\mathcal{I}(a, b) \wedge \mathcal{I}(b, a) = ((1 - a) \wedge (1 - b)) + a \wedge b.$$

Proof We show that

$$\begin{aligned} &(1 - a + b) \wedge (1 - b + a) \wedge 1 \\ &= ((1 - a) \wedge (1 - b)) + a \wedge b. \end{aligned}$$

Suppose that $a \leq b$. Then

$$((1 - a) \wedge (1 - b)) + a \wedge b = 1 - b + a.$$

Clearly, $1 - a + b \geq 1$ and $1 - b + a \leq 1$. Thus $(1 - a + b) \wedge (1 - b + a) \wedge 1 = 1 - b + a$.

The proof of the case for $a \geq b$ is similar. □

Proposition 1.5.7 ([16]) *Let* $\mathcal{I}$ *be the Luckasiewicz implication operator. Then for all* $\mu_A, \mu_B \in \mathcal{FP}(X)$, $E_{\mathcal{I}}(\mu_A, \mu_B) = L(\mu_A, \mu_B)$.

Proof We have that

$$\begin{aligned}
L(\mu_A, \mu_B) &= 1 - \vee\{|\mu_A(x) - \mu_B(x)| \; |x \in X\} \\
&= \wedge\{1 - |\mu_A(x) - \mu_B(x)| \; |x \in X\} \\
&= \wedge\{1 - \mu_A(x) \vee \mu_B(x) + \mu_A(x) \wedge \mu_B(x))|x \in X\} \\
&= \wedge\{(1 - \mu_A(x)) \wedge (1 - \mu_B(x)) + \mu_A(x) \wedge \mu_B(x)|x \in X\}.
\end{aligned}$$

It suffices to show that

$$\mathcal{I}(\mu_A(x), \mu_B(x)) \wedge \mathcal{I}(\mu_B(x), \mu_A(x)) = ((1 - \mu_A(x)) \wedge (1 - \mu_B(x)) + \mu_A(x) \wedge \mu_B(x).$$

However, this holds from Lemma 1.5.6. □

We next consider the interactions between the concept of degree of sameness and fuzzy set theoretical operations.

A fuzzy complement c is called **involutive** if for all $x \in [0, 1]$, $c(c(x)) = x$.

An implication operator $\mathcal{I}$ is called **contrapositive** (with respect to a fuzzy complement c) if $\forall x, y \in [0, 1], \mathcal{I}(x, y) = \mathcal{I}(c(y), c(x))$. We note that the standard complement is involutive.

Proposition 1.5.8 *[166] Let $\mathcal{I}$ be a contrapositive implication operator with respect to an involutive fuzzy complement c. Let μ, ν be fuzzy subsets of X. Then $E_{\mathcal{I}}(\mu, \nu)$ $= E_{\mathcal{I}}(\nu^c, \mu^c)$.*

Proof We have that

$$\begin{aligned}
E_{\mathcal{I}}(\mu, \nu) &= (\wedge\{\mathcal{I}(\mu(x), \nu(x))|x \in X\}) \wedge (\wedge\{\mathcal{I}(\nu(x), \mu(x))|x \in X\}) \\
&= (\wedge\{\mathcal{I}(\nu^c(x), \mu^c(x))|x \in X\}) \wedge (\wedge\{\mathcal{I}(\mu^c(x), \nu^c(x))|x \in X\}) \\
&= E_{\mathcal{I}}(\nu^c, \mu^c).
\end{aligned}$$

□

Proposition 1.5.8 holds for Kleene-Dienes implication operator, $\mathcal{I}(x, y) = (1 - x) \vee y$ and the Early Zadeh implication operator $\mathcal{I}(x, y) = (x \wedge y) \vee (1 - x)$ even though these implication operators are not contrapositive, [166].

1.6 Fuzzy Graph Theory

Let V be a nonempty set. Let $\mathcal{E}$ denote the set of all subsets of V with cardinality 2. Let $E \subseteq \mathcal{E}$. A graph is a pair (V, E). The elements of V are thought of as vertices of the graph and E as the set of edges. For $\{x, y\} \in E$, we let xy denote $\{x, y\}$. Then clearly $xy = yx$.

Most of the following basic definitions are from Rosenfeld's basic work [143].

Definition 1.6.1 Let (V, E) be a graph. Then the pair (σ, μ) is called a **fuzzy subgraph** of (V, E) if σ is a fuzzy subset of V and μ is a fuzzy subset of E such that for all $xy \in E$, $\mu(xy) \leq \sigma(x) \wedge \sigma(y)$.

Definition 1.6.2 Let (σ, μ) be a fuzzy subgraph of the graph (V, E). Then a fuzzy subgraph (τ, ν) of (V, E) is called a **partial fuzzy subgraph** (σ, μ) if $\tau \subseteq \sigma$ and $\nu \subseteq \mu$.

Definition 1.6.3 Let (σ, μ) be a fuzzy subgraph of the graph (V, E). Then a partial fuzzy subgraph (τ, ν) of (σ, μ) is said to **span** (σ, μ) if $\tau = \sigma$. In this case, (τ, ν) is called a **spanning fuzzy subgraph** of (σ, μ).

A **path** P in a fuzzy graph (σ, μ) of a graph (V, E) is a sequence of distinct vertices $x_0, x_1, \ldots, x_n$ (except possibly x_0 and x_n) such that $\mu(x_{i-1}x_i) > 0, i = 1, \ldots, n$. Here n is called the **length** of the path. The consecutive pairs are called the edges of the path. The **diameter** of $x, y \in V$, written diam(x, y), is the length on the longest path joining x and y. The **strength** of P is defined to be $\wedge_{i=1}^{n} \mu(x_{i-1}x_i)$. The **strength of connectedness** between two vertices x and y is defined as the maximum of the strengths of all paths between x and y and is denoted by $\mu^\infty(x, y)$ or CONN(x, y). A **strongest path** joining any two vertices x and y has strength $\mu^\infty(x, y)$. It can be shown that if (τ, ν) is a partial fuzzy subgraph of (σ, μ), then $\nu^\infty \subseteq \mu^\infty$. We call P a **cycle** if $x_0 = x_n$ and $n \geq 3$. Two vertices that are joined by a path are called **connected.** It follows that this notion of connectedness is an equivalence relation. The equivalence classes of vertices under this equivalence relation are called **connected components** of the given fuzzy subgraph. They are its maximal connected partial fuzzy subgraphs.

Let $G = (\sigma, \mu)$ be a fuzzy graph, let x, y be distinct vertices, and let G' be the partial fuzzy subgraph of G obtained by deleting the edge xy. That is, $G' = (\sigma, \mu')$, where $\mu'(x, y) = 0$ and $\mu' = \mu$ for all other pairs. We call xy a **fuzzy bridge** in G if $\mu'^\infty(u, v) < \mu^\infty(u, v)$ for some u, v in σ^*. In words, the deletion of xy reduces the strength of connectedness between some pair of vertices in G. Thus xy is a fuzzy bridge if and only if there exists $u, v \in V$ such that xy is an edge of every strongest path between u and v. That is if xy is a fuzzy bridge of G, then $CONN_G(x, y) = \mu(xy)$ [90]. If the removal of a vertex $w \in \sigma^*$ reduces the strength of connectedness between some pair of vertices in G, then w is called a **fuzzy cutvertex** of G. A fuzzy graph without fuzzy cutvertices is called a **fuzzy block** or simply a **block**.

Definition 1.6.4 ([100]) An edge xy of a fuzzy graph G is called an α-**strong** edge if $\mu(xy) > CONN_{G-xy}(x, y)$, β-**strong** if $\mu(xy) = CONN_{G-xy}(x, y)$, and a δ-**edge** if $\mu(xy) < CONN_{G-xy}(x, y)$. Thus an edge is **strong** if $\mu(xy) \geq CONN_{G-xy}(x, y)$. A path P is called a **strong path** if all of its edges are strong.

G is said to be $\alpha-$saturated, if at least one $\alpha-$strong edge is incident at every vertex $v \in \sigma^*$, and is called $\beta-$saturated, if at least one $\beta-$strong edge is incident at every vertex. G is called **saturated** if it is both α and $\beta-$saturated [93].

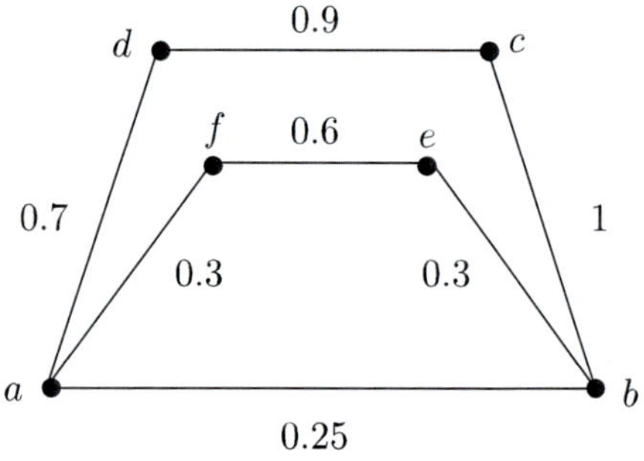

Fig. 1.1 Illustration of paths and their strengths

A fuzzy subgraph $H = (\tau, \nu)$ of $G = (\sigma, \mu)$ is called a **fuzzy cycle**, if $H = (\tau^* \nu^*)$ is a cycle and has more than one weakest edge in ν^*. A connected fuzzy graph $G = (\sigma, \mu)$ is called a **fuzzy tree** if it has a spanning fuzzy subgraph $F = (\sigma, \nu)$ which is a tree, where for all xy not in ν^*, there exists a path from x to y in F, whose strength is more than $\mu(xy)$. Also F is the unique maximum spanning tree of G [90]. A connected fuzzy graph with no fuzzy cycles is a fuzzy tree. A fuzzy graph $G = (\sigma, \mu)$ is called a **fuzzy forest** if each component of G is a fuzzy tree. If there is at most one strongest path between any two vertices of G, then G is a fuzzy forest. A fuzzy graph G is a fuzzy forest if and only if there exists a unique strong path between any two vertices of G.

A **complete fuzzy graph** $G = (\sigma, \mu)$ is one with the property that, for all $x, y \in \sigma^*$, $\mu(xy) = \sigma(x) \wedge \sigma(y)$. A complete fuzzy graph has no fuzzy cutvertices and has at most one fuzzy bridge. If G is a complete fuzzy graph, then for any edge $uv \in \mu^*$, $CONN_G(u, v) = \mu(uv)$ [90].

The **degree** of a vertex $v \in \sigma^*$ is defined as $d(v) = \sum_{u \neq v} \mu(uv)$. The **minimum degree** of G is $\delta(G) = \wedge\{d(v) : v \in \sigma^*\}$ and the **maximum degree** of G is $\Delta(G) = \vee\{d(v) = v \in \sigma^*\}$. The **strong degree** of a vertex $v \in \sigma^*$ is defined as the sum of membership values of all strong edges incident at v. It is denoted by $d_s(v)$. Also if $N_s(v)$ denote the set of all strong neighbors of v, then $d_s(v) = \sum_{u \in N_s(v)} \mu(uv)$.

Example 1.6.5 Consider the fuzzy graph given in Fig. 1.1. with $\sigma^* = \{a, b, c, d, e, f\}$, $\mu(ab) = 0.25$, $\mu(ad) = 0.7$, $\mu(cd) = 0.9$, $\mu(bc) = 1$, $\mu(be) = 0.3$, $\mu(af) = 0.3$ and $\mu(fe) = 0.6$. Here, the path $P_1 = a - d - c - b$ is a strongest path with strength 0.7 and $P_2 = a - f - e - b$ is a strong path with strength 0.25.

The idea of a block in a fuzzy graph was given by Rosenfeld [143]. Excellent studies on blocks were done by Sunitha and Vijayakumar [158] in 2005 and by Mathew and Sunitha [94] in 2010. Characterizations of blocks on at least three vertices, using strongest paths and strongest strong paths can be seen in these works.

Theorem 1.6.6 ([90]) *The following statements are equivalent for a connected fuzzy graph $G = (\sigma, \mu)$.*

1. G is a block.

2. *Any two vertices u and v such that uv is not a fuzzy bridge are joined by two internally disjoint strongest paths.*
3. *For every three distinct vertices of G, there is a strongest path joining any two of them not containing the third.*

In 2003, Bhutani and Rosenfeld [23–25] examined cycles and categorized them as **locamin cycles** *and* **multimin cycles**. *If every vertex of the cycle* $C = (\rho, \nu)$ *lies on a weakest edge, then C is called a locamin cycle, and if C has more than one weakest edge, then C is called a multimin cycle. A fuzzy cycle is always a multimin cycle.*

Definition 1.6.7 ([99]) A cycle C in a fuzzy graph G is called a **strongest strong cycle**(SSC) if C is the union of two strongest strong uv paths for every pair of vertices u and v in C except when uv is a fuzzy bridge of G in C.

If G^* is a cycle, then the concepts of strongest strong cycle and locamin cycle coincide and it becomes a block.

Theorem 1.6.8 ([90]) *Let* $G = (\sigma, \mu)$ *a fuzzy graph such that* G^* *is a cycle. Then the following are equivalent.*

1. *G is a block.*
2. *G is an SSC.*
3. *G is a locamin cycle.*

Definition 1.6.9 ([171]) A **disconnection** of a fuzzy graph $G = (\sigma, \mu)$ is a set $D \subset \sigma^*$ whose removal results in a disconnected fuzzy graph or $G - D$ becomes trivial, where $G - D$ represents the fuzzy graph with all vertices in D, deleted. The weight of D is defined as $\sum_{v \in D} \min\{\mu(vu) | \mu(vu) \neq 0\}$. The **vertex connectivity**, $\Omega(G)$ of G is the minimum weight of a disconnection.

Let G be a fuzzy graph and $\{V_1, V_2\}$ be a partition of its vertex set. The set of edges joining vertices of V_1 and vertices of V_2 is called a **cut-set** of G, denoted by (V_1, V_2) relative to the partition $\{V_1, V_2\}$. The weight of the cut-set (V_1, V_2) is defined as $\sum_{u \in V_1, v \in V_2} \mu(uv)$. Let G be a fuzzy graph. The **edge connectivity** of G denoted by $\lambda(G)$ is defined to be the minimum weight of cut-sets of G [171].

Definition 1.6.10 ([98]) A set of vertices $X = \{v_1, v_2, \ldots, v_m\}$ is said to be a **fuzzy vertex cut** if either, $CONN_{G-X}(x, y) < CONN_G(x, y)$ for some pair of vertices $x, y \in \sigma^*$ such that both $x, y \neq v_i$ for $i = 1, 2, \ldots, m$ or $G - X$ is trivial. The **strong weight** $s(X)$ of X is,

$$s(X) = \sum_{x \in X} \mu(xy),$$

where $\mu(xy)$ is the minimum of the weights of strong edges incident on x.

Definition 1.6.11 ([98]) The **fuzzy vertex connectivity** of a connected fuzzy graph G is defined as the minimum strong weight of fuzzy vertex cuts of G. It is denoted by $\kappa(G)$.

Definition 1.6.12 ([98]) Let G be a connected fuzzy graph and $t \in (0, \infty)$; G is called *t-connected* if $\kappa(G) \geq t$.

Definition 1.6.13 ([98]) Let $G = (\sigma, \mu)$ be a fuzzy graph. A set of strong edges $E = \{e_1, e_2, \ldots, e_n\}$ with $e_i = u_i v_i, i = 1, 2, \ldots, n$ is said to be a **fuzzy edge cut** or **fuzzy arc cut (FAC)** if either $CONN_{G-X}(x, y) < CONN_G(x, y)$ for some pair of vertices $x, y \in \sigma^*$ with at least one of x or y different from both u_i and v_i, $i = 1, 2, \ldots, n$, or $G - E$ is disconnected. The **strong weight** of a fuzzy edge cut E is defined as $s'(E) = \sum_{e_i \in E} \mu(e_i)$.

Definition 1.6.14 ([98]) The **fuzzy edge connectivity** $\kappa'(G)$ of a connected fuzzy graph G is defined to be the minimum strong weight of fuzzy edge cuts of G. Let G be a connected fuzzy graph and $t \in (0, \infty)$. G is called t-**edge connected** if $\kappa'(G) \geq t$.

A generalization of Menger's theorem for fuzzy graphs was given by Mathew and Sunitha [97] in 2013, and they accomplished it using the concept of **strength reducing sets**. A set $S \subseteq \sigma^*$ is a $u - v$ strength reducing set of vertices, where u, v are two distinct vertices in G such that uv is not strong, if the removal of S from G reduces the strength of connectedness between u and v. A Necessary and sufficient condition for a set of vertices to be a strength reducing set is given below.

Theorem 1.6.15 ([97]) *Let $G = (\sigma, \mu)$ be a connected fuzzy graph and u, v, any two vertices in G such that uv is not strong. Then a set S of vertices in G is a $u - v$ strength reducing set if and only if every strongest path from u to v contains at least one vertex of S.*

Next we have generalization of vertex version of Menger's Theorem. It relates a $u - v$ strength reducing set to the number of internally disjoint strongest $u - v$ paths in G.

Theorem 1.6.16 ([97]) *Let $G = (\sigma, \mu)$ be a fuzzy graph. For any two vertices $u, v \in \sigma^*$ such that uv is not strong, the maximum number of internally disjoint strongest $u - v$ paths in G is equal to the number of vertices in a minimal $u - v$ strength reducing set.*

Theorem 1.6.16 and Definition 1.6.17 will be used in Chap. 8.

Definition 1.6.17 ([99]) Let $G = (\sigma, \mu)$ be a fuzzy graph. Then for any two nodes u and v of G, there associated a set say $\theta(u, v)$ called the θ-**evaluation** of u and v and is defined as $\theta(u, v) = \{\alpha : \alpha \in (0, 1]\}$ where α is the strength of a strong cycle passing through both u and v. Max$\{\alpha : \alpha \in \theta(u, v); u, v \in \sigma^*\}$ is defined as the **cycle connectivity** between u and v in $G : (\sigma, \mu)$ and is denoted by $C^G_{u,v}$. If $\theta(u, v) = 0$ for some pair of nodes u and v, we define the cycle connectivity between u and v to be 0.

Definition 1.6.17 has been used extensively in the literature following the introduction of cycle connectivity index of a fuzzy graph, which will be discussed in Chap. 8.

Recently two useful connectivity indices namely Connectivity Index and Wiener Index were introduced to fuzzy graph theory, whose definitions are given below.

Definition 1.6.18 ([28]) Let $G = (\sigma, \mu)$ be a fuzzy graph. The **Connectivity Index** (CI) of G is defined by

$$CI(G) = \sum_{u,v \in \sigma*} \sigma(u)\sigma(v)CONN_G(u, v)$$

where $CONN_G(u, v)$ is the strength of connectedness between u and v.

Definition 1.6.19 ([29]) Let $G = (\sigma, \mu)$ be a fuzzy graph. The $Wiener Index$ (WI) of G is defined by

$$WI(G) = \sum_{u,v \in \sigma^*} \sigma(u)\sigma(v)d_s(u, v),$$

where $d_s(u, v)$ is the minimum sum of weights of geodesics from u to v.

Much of the works on these indices can be found in [110, 118]. The relationship between these indices and cyclic connectivity index will be discussed in Chap. 8.

Chapter 2
Sustainable Development Goals

The following is taken from [36]. The overwhelming majority of climate scientists agree that human activities, especially the burning of fossil fuels (coal, oil, and gas), are responsible for most of the climate change currently being observed. Scientists have been taking widespread measurements of Earth's surface temperature since around 1880. Temperatures are recorded by thermometers at many thousands of locations, both on the land and over the oceans. Research groups work carefully to make sure the data aren't skewed by such things as changes in the instruments taking the measurements or by other factors that affect local temperature, such as additional heat that has come from the gradual growth of cities. These analyses all show that the Earth's surface temperature has increased by more than 1.4 °F (0.8 °C) over the past 100 years, with much of the increase over the past 35 years.

Humans are causing greenhouse gas concentrations to increase. (Greenhouse gases are gases in Earth's atmosphere that trap heat. They let sunlight pass through the atmosphere, but they prevent the heat from leaving the atmosphere. The main greenhouse gases are water vapor, carbon dioxide, methane, ozone, nitrous acid, chlorofluorocarbons.) Once humans began digging up long-buried forms of carbon such as coal and burning them for energy, additional CO_2 began to be released into the atmosphere much more rapidly than in the natural carbon cycle. Other human activities, such as cement production and cutting down and burning of forests (deforestation), also add CO_2 to the atmosphere. Human activities have increased the atmospheric concentrations of other important greenhouse gases as well. Methane, which is produced by burning of fossil fuels, the raising of livestock, the decay of landfill wastes, the production and transport of natural gas, and other activities, increases sharply through the 1980s before starting to level off at about two-and-half times its preindustrial level. Nitrous oxide has increased by approximately 25% since 1750, mainly as a result of agricultural fertilizer use, but also from fossil fuel burning and certain industrial processes.

The following is taken from [131]. Time is running out to limit global temperature rise to well below 2 °C, aiming for 1.5 °C. To accomplish this limit, emissions must

J. N. Mordeson and S. Mathew, *Mathematics of Uncertainty for Coping with World Challenges*, Studies in Systems, Decision and Control 353,
https://doi.org/10.1007/978-3-030-68684-0_2

peak as soon as possible and decline sharply until 2050. However, the world is not on that path. Future Earth and the Earth League collated the most up-to-date science since 2017, on the drivers and impacts of climate change. They make the case that achieving the Paris Agreement is not only necessary, but possible. Meeting the goals of the Agreement entails conserving and sustainably using nature and social paradigms, and compels new governance models and improved capacity to adapt to climate change.

Growing climate impacts increase the risks of critical tipping points that, if crossed, lead to far-reaching, and/or irreversible consequences for the stability of life on Earth. (A tipping point in the climate system is a threshold that, when exceeded, can lead to large changes in the state of the system. As an example, self-reinforcing feedbacks in the carbon cycle and planetary reflectivity could trigger a cascading set of tipping points that lead the world into a hothouse climate.) It is also now known that there are large differences in climate change impacts between 1.5 °C and 2 °C.

Key processes that currently keep the climate stable are weakening, risking the establishment of feedback loops (e.g. loss of Arctic sea ice or forest dieback) that could hinder efforts to stabilize the climate, even if emissions are reduced.

There is a growing recognition that the risk of increased warming has been underestimated, as climate impacts are hitting harder and sooner than anticipated. As climate change intensifies, cities are particularly vulnerable to impacts such as heat stress and can also play a key role in reducing emissions locally and globally.

Strategies for mitigation and for upscaling adaptive risk management are necessary going forward. Neither is adequate in isolation given the pace of climate change and magnitude of its impacts. Only immediate and all-inclusive action will enable us to meet the Paris Agreement target of well below 2 °C. Action must encompass deep decarbonization complemented by ambitious policy measures, protection and enhancement of carbon sinks and biodiversity, and efforts to remove CO_2 from the atmosphere.

We compare the rankings of the countries in their achievement of the SDGs by using the similarity measures discussed in Chap. 1. We take the average of the three rankings using the AHP, Guiasu, and Yen method to determine one ranking. We also take the average of the three averages determined by the Stakeholder method to obtain one ranking. We then determine the ranking obtained in [149] for our third ranking. We make our comparisons in Sect. 2.2 of these three rankings.

2.1 Average of Rankings

All member states of the United Nations adopted Agenda 2030 and the Sustainable Development Goals (SDGs) in 2015. The 17 SDGs describe a universal agenda that applies to and must be implemented by all countries, both developed and developing, [148]. It is stated in [148] that sound metrics and data are critical for turning the SDGs into practical tools for problem-solving by (a) mobilizing governments, academia, civil society, and business, (b) providing a report card to track progress and ensure accountability, and (c) serving as a management tool for the transformation needed to achieve the SDGs by 2030.

In [113] the metrics and data provided in [133, 149] were taken and transformed into a fuzzy logic setting. This allowed for the analysis of the results in [133, 149] by using mathematics of uncertainty. In [113], the Analytic Hierarchy Process (AHP), the Guiasu method, and the Yen method were used to determine which countries are, in the opinion of experts, the best in achieving sustainability. This was accomplished by constructing linear equations involving the 17 SDGs as independent variables and a measure of success as the dependent variable. The ranking of the countries was based on the philosophy that for a goal or target to score highly, it must meet all of certain criteria.

The 17 SDGs are G_1 : No Poverty, G_2 : Zero Hunger, G_3 : Good health and Well Being, G_4 : Quality Education, G_5 : Gender Equality, G_6 : Clean Water and Sanitation, G_7 : Affordable and Clean Energy, G_8 : Decent Work and Economic Growth, G_9 : Industry, Innovation, and Infrastructure, G_{10} : Reduced Inequalities, G_{11} : Sustainable Cities and Communities, G_{12} : Responsible Construction and Production, G_{13} : Climate Action, G_{14} : Life Below Water, G_{15} : Life on Land, G_{16} : Peace Justice, and Strong Institutions, G_{17} : Partnerships and Goals. These SDGs are discussed in more detail in [149] and Appendix 2.1.

In [133], Stakeholder Forum created a methodology or analytical tool to enable relative scores to be assigned to each of the different targets and goals according to their significance in different contexts. The method uses a number of assessors to assign their own independent scores of the significance of each of the proposed targets in the implementation context in question, according to three separate criteria. The three criteria proposed are applicability, implementability, and the transformational impact (both in the country concerned and for the world as a whole, [133]. The assessors' scores are then aggregated and averaged to give an overall score for each target, and then combined to give an average score for each goal. The methodology is described in more detail in Chap. 1.

Goal 17 and the targets within other goals that are specifically directed towards international cooperation and development assistance responsibilities of developed countries are excluded from the analysis in [133] and consequently here also.

We next present the equations used in [113] to rank the countries with respect to their achievement of the SDGs. In order to be in keeping with the approach in [133], the development of the equations relied on the concept of a t-norm. This was because the approach in [133] used the philosophy that for a goal or target to score highly, it must meet all three criteria mentioned above.

AHP equation:

$$\begin{aligned} G = {} & 0.03G_1 + 0.02G_2 + 0.03G_3 + 0.04G_4 + 0.04G_5 + 0.05G_6 + 0.12G_7 \\ & + 0.06G_8 + 0.04G_9 + 0.07G_{10} + 0.05G_{11} + 0.12G_{12} + 0.13G_{13} \\ & + 0.08G_{14} + 0.05G_{15} + 0.05G_{16}. \end{aligned}$$

Guiasu equation:

$$G' = 0.03G_1 + 0.02G_2 + 0.02G_3 + 0.04G_4 + 0.04G_5 + 0.05G_6 + 0.13G_7 + 0.05G_8 + 0.04G_9 + 0.07G_{10} + 0.05G_{11} + 0.13G_{12} + 0.14G_{13} + 0.09G_{14} + 0.05G_{15} + 0.05G_{16}.$$

Yen equation:

$$G'' = 0.03G_1 + 0.03G_2 + 0.03G_3 + 0.05G_4 + 0.05G_5 + 0.04G_6 + 0.13G_7 + 0.05G_8 + 0.05G_9 + 0.07G_{10} + 0.05G_{11} + 0.13G_{12} + 0.14G_{13} + 0.05G_{14} + 0.05G_{15} + 0.05G_{16}.$$

In [133], Stakeholder Forum created a methodology to enable relative scores to be assigned to each of the different targets and goals according top their difference significance in different contexts. The method has been discussed in Chap. 1. The following table is determined from [133].

The ranking of the SDGs in the following table are by level of transformational challenge in developed countries as determined in [133]. The overall marks are from Table 2.1 in [[133], p. 6].

Goal	Overall mark for goal	weight
G_1	1.8	0.03
G_2	2.3	0.04
G_3	1.5	0.03
G_4	2.5	0.05
G_5	2.2	0.04
G_6	2.5	0.05
G_7	6.4	0.12
G_8	2.7	0.05
G_9	2.1	0.04
G_{10}	3.6	0.07
G_{11}	2.6	0.05
G_{12}	6.3	0.12
G_{13}	6.4	0.13
G_{14}	4.4	0.08
G_{15}	2.7	0.05
G_{16}	2.7	0.05
Column Sum	53.4	

The weights in the table are the coefficients in the following equation. They are determined by dividing the entries in the overall mark column by 53.4.

Stakeholder equation:

$$\begin{aligned} SDG = {} & 0.03G_1 + 0.04G_2 + 0.03G_3 + 0.05G_4 + 0.04G_5 + 0.05G_6 \\ & + 0.12G_7 + 0.05G_8 + 0.04G_9 + 0.07G_{10} + 0.05G_{11} + 0.12G_{12} \\ & + 0.13G_{13} + 0.08G_{14} + 0.05G_{15} + 0.05G_{16}. \end{aligned}$$

In [113], the averages of the rankings mentioned above were determined. These averages appear in the first two columns of the following tables. Based on these averages, we first provide the overall rank of the country as determined in [113]. We call the method in [149], the SD Report. We then use this ranking to determine the rank of the country in their particular region. The tables are from [[113], pp. 110–116] and [149].

The Scandinavian countries do well in achieving the sustainable development goals with respect to other countries in the region. However the OECD countries are not on track for achieving the SDGs. Compared to the rest of the world, OECD countries perform better on goals related to socio-economic outcomes and basic access to infrastructure, [149] (Tables 2.1, 2.2, 2.3, 2.4, 2.5, 2.6 and 2.7).

SDG challenges vary greatly across countries in East and South Asia due in part to the difference in their size and economic development, [149]. Overall, best performance is obtained on SDG 1, SDG 4, and SDG 7.

Belarus, Croatia, Moldova, and Kyrgz Rep. rank high within the region. In general, countries in Eastern Europe and Central Asia obtain their best performance on SDG 1 and SDG 7, [149].

Costa Rica ranks the highest in all three rankings. Latin America and Caribbean countries perform best on SDG 1 and SDG 7. Compared to other parts of the world, further efforts are needed to reduce income and wealth inequalities, [149].

The performance of Middle East and North African countries varies greatly across countries, [149].

Cabo Verde and Gabon rank highly in all three rankings within the region. All sub-Saharan African countries face major challenges in achieving the SDGs. Performance on socio-economic goals and basic access to services and infrastructure are poor compared to other regions in the world, [149].

There is a relative lack of comparable data available for the regions to benchmark these small island countries against the rest of the world, [149].

Table 2.1 OECD combined ranks

OECD rankings			
Country	AHP, Guiasu, Yen rank	Stakeholder rank	SD report rank (Regional)
Australia	36	29	38 /32
Austria	5	8.5	5 /5
Belgium	24.5	21	16 /16
Canada	27	11	20 /20
Chile	26	30	31 /28
Czech Rep.	1	10	7 /7
Denmark	2	2	1/1
Estonia	7	14	10 /10
Finland	9	1	3 /3
France	7	13	4 /4
Germany	19	22	6 /6
Greece	34	32	50 /34
Hungary	3	18	25 /23
Iceland	18	7	14 /14
Ireland	12	16.5	19 /19
Israel	30	33	49 /33
Italy	28	24	30 /27
Japan	17	16.5	15 /15
Korea Rep.	14.5	25	18 /18
Latvia	16	28	24 /22
Lithuania	23	31	32 /29
Luxembourg	31	26	34 30
Mexico	32.5	35	78 /35
Netherlands	10.5	12	9 /9
New Zealand	10.5	8.5	11 /11
Norway	29	4	8 /8
Poland	22	27	29 /26
Portugal	24.5	15	26 /24
Slovak Rep.	14.5	19	27 /25
Slovenia	7	6	12 /12
Spain	21	20	21 /21
Sweden	4	3	2 /2
Switzerland	13	5	17 /17
Turkey	35	36	79 /36
U. K	20	23	13 /13
United States	32.5	34	35 /31

Table 2.2 East and South Asia combined ranks

East and South Asia			
Country	AHP, Guiasu, Yen rank	Stakeholder rank	SD report rank (Regional)
Bangladesh	13	12	116 /17
Bhutan	4	3	84 /7
Brunei Darussalam	11.5	8.5	
Cambodia	17	19	112 /15
China	2	10	39 /1
India	16	17	115 /16
Indonesia	9	11	102 /11
Korea, Dem. Rep.			
Lao PDR	11.5	8.5	111 /14
Malaysia	6	5.5	68 /6
Maldives	5	2	47 /3
Mongolia	18	18	100/10
Myanmar	15	13	110 /13
Nepal	7	5.5	103 /12
Pakistan	19	16	130 /18
Philippines	8	14.5	97 /9
Singapore	14	1	66 /5
Sri Lanka	10	7	93 /8
Thailand	1	14.5	40 /2
Timor Leste			
Vietnam	3	4	54 /4

Table 2.3 Eastern Europe and Central Asia combined ranks

Eastern Europe and Central Asia			
Country	AHP, Guiasu, Yen rank	Stakeholder rank	SD report rank (Regional)
Afghanistan	23		153 /23
Albania	13	14	60 /13
Andorra			
Armenia	7.5	3	75 /19
Azerbaijan	14	13	59 /12
Belarus	1	7.5	23 /2
Boznia and Herzegovina	20	16.5	69 /15
Bulgaria	11	9	36 /4
Croatia	3	7.5	22 /1
Cyprus	18	19	61 /14
Georgia	16	15	73 /18
Kazakhstan	19	20	77 /20
Kyrgz Rep.	4	2	48 /9
Liecherstan			
Malta	7.5	5.5	28 /3
Moldova	2	1	37 /5
Monaco			
Montenegro	21	21	87 /21
North Macedonia	15	18	70 /16
Romania	12	12	42 /7
Russian Federation	17	16.5	55 /11
San Marino			
Serbia	9.5	10.5	44 /8
Tajikistan	6	5.5	71 /17
Turkmenistan	22		101 /22
Ukraine	9.5	4	41 /6
Uzbekistan	5	10.5	52 /10

Table 2.4 Latin America and the Caribbean combined ranks

Latin America and the Caribbean			
Country	AHP, Guiasu, Yen rank	Stakeholder rank	SD report rank (Regional)
Antigua and Barbuda			
Argentina	8	14	45 /3
Bahamas, The Barbados			
Belize	16.5	17	109 /20
Bolivia	7	16	80 /11
Brazil	9.5	8	57 /7
Columbia	5.5	3	67 /9
Costa Rica	1	1	33 /1
Cuba	13	4	56 /6
Dominica			
Dominican Rep.	4	6	64 /8
Ecuador	2	2	46 /4
El Salvador	16.5	10	89 /16
Grenada			
Guatamala	21	22	122 /22
Guyana	22	19.5	114 /21
Haiti	23	23	156 /23
Honduras	18	21	107 /18
Jamaica	14	15	74 /10
Nicaragua	11	11.5	82 /12
Panama	15	7	90 /17
Paraguay	12	13	86 /14
Peru	3 /3	11.5	51 /5
St. Kitts and Nevis			
St. Lucia			
St. Vincent and the Grenadines			
Suriname	9.5	9	88 /15
Trinidad and Tobago	20	18	85 /13
Uruguay	5.5	5	43 /2
Venezuelia, RB	19	19.5	108 /19

Table 2.5 Middle East and North Africa combined ranks

Middle East and North Africa			
Country	AHP, Guiasu, Yen rank	Stakeholder rank	SD report rank (Regional)
Algeria	1	1	53 /1
Bahrain	10	2	76 /6
Egypt, Arab Rep	6	3	92 /10
Iran, Islamic Rep.	2	6	58 /2
Iraq	9	9.5	117 /14
Jordan	5	4	81 7
Kuwait	12	15	106 /13
Lebanon	7	14	94 /11
Libya			
Morocco	4	11	72 /5
Oman	8	9.5	83 /8
Qatar	15	7	91 /9
Saudi Arabia	13	13	98 /12
Syrian Arab Republic	11	12	123 /15
Tunisia	3	5	63 /3
United Arab Emirates	14	8	65 /4
Yemen, Rep.	16	16	133 /16

Table 2.6 Sub-Saharan Africa combined ranks

Sub-Saharan Africa			
Country	AHP, Guiasu, Yen rank	Stakeholder rank	SD report rank (Regional)
Angola	25	42.5	149 /32
Benin	34.5	16	151 /34
Botswana	10	32	120 /8
Burkino Faso	31	12	141 /25
Burundi	6	4	145 /28
Cabo Verde	2.5	2	96 /2
Cameroon	16	24	127 /13
Central African Rep.	42	13	162 /43
Chad	43	35	161 /42
Comoros	29.5	41	137 /21
Congo Democratic Rep.	40	29	160 /41
Congo Rep.	19	34	132 /17
Cote d'Ivoire	13	14	129 /15

(continued)

Table 2.6 (continued)

Sub-Saharan Africa

Country	AHP, Guiasu, Yen rank	Stakeholder rank	SD report rank (Regional)
Djibouti	37	42.5	148 /31
Equatoriaj Guinea			
Erita			
Eswatini	29.5	40	142 /26
Ethiopia	17	7	135 /19
Gabon	1	3	99 /3
Gambia, The	23	30	131 /16
Ghana	4.5	31	104 /4
Guinea	24	17	138 /22
Guinea-Bissau			
Kenya	14	27.5	125 /11
Lesotho	41	38	150 /33
Liberia	34.5	11	157 /38
Madagascar	38	37	158 /39
Malawi	4.5	25	146 /29
Mali	28	6	152 /35
Mauritania	27	33	134 /18
Mauritius	9	9	105 /5
Mozambique	18	19.5	136 /20
Nambia	12	36	119 /7
Niger	36	19.5	154 /36
Nigeria	39	23	159 /40
Rwanda	22	8	126 /12
Sao Tome and Principe	2.5	10	95 /1
Senegal	11	39	124 /10
Seychelles			
Sierra Leone	33	18	155 /37
Somolia			
South Africa	7	26	113 /6
South Sudan			
Sudan	20	27.5	147 /30
Tanzania	15	5	128 /14
Togo	32	21.5	144 /27
Uganda	26	15	140 /24
Zambia	21	21.5	139 /23
Zimbabwe	8	1	121 /9

Table 2.7 Oceania combined ranks

Oceania			
Country	AHP, Guiasu, Yen rank	Stakeholder rank	SD report rank (Regional)
Fiji	1	1	62 /1
Kiribati			
Marshall Islands			
Micronesia, Fed. Sts.			
Nauru			
Palau			
Papua New Guinea	3	2.5	143 /3
Samoa			
Solomon Islands			
Tonga			
Tuvalu			
Vanuatu	2	2.5	118 /2

2.2 Similarity Rankings

We next determine the similarity of various rankings for each region, except Oceania. There isn't sufficient data for Oceania.

In the following, when a ranking has missing data for a country, we delete the country for a different ranking even if it has data for the ranking. After the deletion, we reranked.

Some of the following results concerning the fuzzy similarity measure M are from [113].

Let μ denote the ranking for the AHP, Guiasu, Yen method, ν the ranking for the stakeholder, and ρ for the SD Report ranking. We first consider the similarity of the SD Report rankings and the Stakeholder ranking for OECD countries. Let $C_1, \ldots, C_{36}$ denote the countries making up the OECD region. We use the similarity operation

$$M(\nu, \rho) = \frac{\sum_{i=i}^{36} \nu(C_i) \wedge \rho(C_i)}{\sum_{i=i}^{36} \nu(C_i) \vee \rho(C_i)}.$$

It follows that $M(\nu, \rho) = 588/744 = 0.790$. By Eq. (1.4.4), the smallest possible similarity measure for 36 countries with no ties in the ranking can be obtained by substituting 36 for n in the formula $\frac{n+2}{3n+2}$. One obtains $38/110 = 0.345$. We have that $\frac{M(\nu,\rho)-0.345}{1-0.345} = \frac{0.790-0.345}{1-0.345} = 0.679$.

We also obtain $M(\mu, \rho) = 582/750 = 0.776$ and $M(\mu, \nu) = 561.5/770.5 = 0.729$.

We also have that $\frac{M(\mu,\rho)-0.345}{1-0.345} = \frac{0.776-0.345}{1-0.345} = 0.658$ and $\frac{M(\mu,\nu)-0.345}{1-0.345} = \frac{0.729-0.345}{1-0.345} = 0.586$.

East and South Asia

$M(\nu, \rho) = 137/205 = 0.668$. We find that $M(\mu, \rho) = 149/193 = 0.772$. The smallest the ranking can be with no ties is $\frac{n+2}{3n+2} = \frac{18+2}{54+2} = \frac{20}{56} = 0.357$ since $n = 18$. We also have that $M(\mu, \nu) = 156/224 = 0.696$. By Eq. (1.4.5), the smallest ranking with no ties is $\frac{n+1}{3n-1} = \frac{19+1}{57-1} = \frac{20}{56} = 0.357$ since here $n = 19$.

We have that $\frac{M(\nu,\rho)-0.357}{1-0.357} = \frac{0.668-0.357}{1-0.357} = 0.475$, $\frac{M(\mu,\nu)-0.357}{1-0.357} = \frac{0.696-0.357}{1-0.357} = 0.518$, and $\frac{M(\mu,\rho)-0.357}{1-0.357} = \frac{0.772-0.357}{1-0.356} = 0.634$.

Eastern Europe and Central Asia

We have that $M(\mu, \nu) = \frac{207.5}{254.5} = 0.815$ with $\frac{n+1}{3n-1} = \frac{21+1}{63-1} = \frac{22}{62} = 0.355$ and $M(\nu, \rho) = \frac{187.5}{274.5} = 0.683$ with $\frac{n+1}{3n-1} = \frac{21+1}{63-1} = \frac{22}{62} = 0.355$ since $n = 21$. We found that $M(\mu, \rho) = \frac{235.5}{316.5} = 0.744$. Here we have 23 countries and $\frac{n+1}{3n-1} = \frac{23+1}{69-1} = \frac{24}{68} = 0.353$.

We have that $\frac{M(\mu,\nu)-0.355}{1-0.355} = \frac{0.815-0.355}{1-0.355} = 0.713$, $\frac{M(\nu,\rho)-0.355}{1-0.355} = \frac{0.683-0.355}{1-0.355} = 0.509$, and $\frac{M(\mu,\rho)-0.353}{10.353} = \frac{0.744-0.353}{1-0.353} = 0.605$.

Latin America and the Caribbean

We have that $M(\mu, \nu) = \frac{243}{309} = 0.786$, $M(\mu, \rho) = \frac{245.5}{306.5} = 0.801$, and $M(\nu, \rho) = \frac{236}{316} = 0.747$. Here we have 23 countries and $\frac{n+1}{3n-1} = \frac{23+1}{69-1} = \frac{24}{68} = 0.353$.

We have that $\frac{M(\mu,\nu)-0.353}{1-0.353} = \frac{0.786-0.353}{1-0.353} = 0.669$, $\frac{M(\mu,\rho)-0.353}{1-0.353} = \frac{0.801-0.353}{1-0.353} = 0.692$, and $\frac{M(\nu,\rho)-0.353}{1-0.353} = \frac{0.747-0.353}{1-0.353} = 0.609$.

Middle East and North Africa

We have that $M(\mu, \nu) = \frac{110}{162} = 0.679$, $M(\mu, \rho) = \frac{115}{157} = 0.732$, and $M(\nu, \rho) = \frac{112.5}{159.5} = 0.705$. Here we have 16 countries and $\frac{n+2}{3n+3} = \frac{16+2}{48+2} = \frac{18}{50} = 0.360$.

We have that $\frac{M(\mu,\nu)-0.360}{1-0.360} = \frac{0.679-0.360}{1-0.360} = 0.498$, $\frac{M(\mu,\rho)-0.360}{1-0.360} = \frac{0.732-0.360}{1-0.360} = 0.581$, and $\frac{M(\nu,\rho)-0.360}{1-0.360} = \frac{0.705-0.360}{1-0.360} = 0.539$.

Sub-Saharan Africa

We have that $M(\mu, \nu) = \frac{691.5}{1200.5} = 0.576$, $M(\mu, \rho) = \frac{853}{1039} = 0.821$, and $M(\nu, \rho) = \frac{669.5}{1222.5} = 0.548$. Here we have 43 countries and $\frac{n+1}{3n-1} = \frac{43+1}{129-1} = \frac{44}{128} = 0.344$.

We have that $\frac{M(\mu,\nu)-0.344}{1-0.344} = \frac{0.576-0.344}{1-0.344} = 0.354$, $\frac{M(\mu,\rho)-0.344}{1-0.344} = \frac{0.821-0.344}{1-0.344} = 0.727$, and $\frac{M(\nu,\rho)-0.344}{1-0.344} = \frac{0.548-0.344}{1-0.344} = 0.311$.

We next consider the fuzzy similarity measure S.

OECD

Let μ denote the ranking for the AHP, Guiasu, Yen method, ν the ranking for the stakeholder, and ρ for the SD Report ranking. We first consider the similarity of the SD

Report rankings and the Stakeholder ranking for OECD countries. Let $C_1, \ldots, C_{36}$ denote the countries making up the OECD region. We use the similarity operation

$$S(\nu, \rho) = 1 - \frac{\sum_{i=1}^{36} |\nu(C_i) - \rho(C_i)|}{\sum_{i=1}^{36} \nu(C_i) + \rho(C_i)}.$$

It follows that $S(\nu, \rho) = 1 - 156/1333 = 0.883$. By Eq. (1.4.6), the smallest possible similarity measure for 36 countries with no ties in the ranking can be obtained by substituting 36 for n in the formula $\frac{n/2+2}{n+1}$. One obtains $20/37 = 0.541$.

We also obtain $S(\mu, \rho)=1 - 168/1332 = 0.874$ and $S(\mu, \nu) = 1 - 209/1332 = 0.843$.

We have that $\frac{S(\nu,\rho)-0.541}{1-0.541} = \frac{0.883-0.541}{1-0.541} = 0.745$, $\frac{S(\mu,\rho)-0.541}{1-0.541} = \frac{0.874-0.541}{1-0.541} = 0.725$, and $\frac{S(\mu,\nu)-0.541}{1-0.541} = \frac{0.843-0.541}{1-0.541} = 0.658$.

East and South Asia

$S(\nu, \rho) = 1 - 68/342 = 0.801$. We find that $S(\mu, \rho) = 1 - 44/342 = 0.871$. The smallest the ranking can be with no ties is $\frac{n/2+1}{n+1} = \frac{9+1}{18+1} = \frac{10}{19} = 0.526$ since $n = 18$. We also have that $S(\mu, \nu) = 1 - 60/342 = 0.825$. Brunei Darussiam has sufficient data for μ and ν. Adding Brunei Darussiam back in yielded $S(\mu, \nu) = 1 - 68/380 = 0.821$ and since $n = 19$ in this case $\frac{1}{2} + \frac{1}{2n} = \frac{1}{2} + \frac{1}{38} = 0.526$.

We have that $\frac{S(\nu,\rho)-0.526}{1-0.526} = \frac{0.801-0.526}{1-0.526} = 0.580$, $\frac{S(\mu,\rho)-0.526}{1-0.526} = \frac{0.871-0.526}{1-0.526} = 0.728$, and $\frac{S(\mu,\nu)-0.526}{1-0.526} = \frac{821-0.526}{1-0.526} = 0.622$.

Eastern Europe and Central Asia

We have that $S(\mu, \nu)=1 - 47/462=0.898$ with $\frac{1}{2} + \frac{1}{42} = \frac{22}{62} = 0.524$ and $S(\nu, \rho) = 1 - 87/462 = 0.812$ and $S(\mu, \rho) = 1 - 81/452 = 0.821$. Since $n = 21$, $\frac{1}{2} + \frac{1}{2n} = \frac{1}{2} + \frac{1}{42} = 0.524$. Afghanistan and Turkmenistan have sufficient data for μ and ρ. Adding Afghanistan and Turkmenistan back in yielded $S(\mu, \rho) = 1 - 81/542 = 0.851$ and since $n = 23$, we have $\frac{1}{2} + \frac{1}{2n} = \frac{1}{2} + \frac{1}{46} = 0.522$.

We have that $\frac{S(\mu,\nu)-0.524}{1-0.524} = \frac{0.898-0.524}{1-0.524} = 0.788$, $\frac{S(\nu,\rho)-0.524}{1-0.524} = \frac{0.812-0.524}{1-0.524} = 0.605$, and $\frac{S(\mu,\rho)-0.522}{1-0.522} = \frac{0.851-0.522}{1-0.522} = 0.688$.

Latin America and the Caribbean

We have that $S(\mu, \nu) = 1 - 66/552 = 0.880$, $S(\mu, \rho) = 1 - 61/571 = 0.893$, and $S(\nu, \rho) = 80/552 = 0.855$. Here we have 23 countries and $\frac{1}{2} + \frac{1}{2n} = \frac{1}{2} + \frac{1}{46} = 0.522$.

We have that $\frac{S(\mu,\nu)-0.522}{1-0.522} = \frac{0.880-0.522}{1-0.522} = 0.749$, $\frac{S(\mu,\rho)-0.522}{1-0.522} = \frac{0.893-0.522}{1-0.522} = 0.776$, and $\frac{S(\nu,\rho)-0.522}{1-0.522} = \frac{0.855-0.522}{1-0.522} = 0.697$.

Middle East and North Africa

We have that $S(\mu, \nu) = 1 - 52/272 = 0.809$, $S(\mu, \rho) = 1 - 42/272 = 0.846$, and $S(\nu, \rho) = 1 - 47/272 = 0.827$. Here we have 16 countries and $\frac{n/2+1}{n+1} = \frac{8+1}{16+1} = 0.529$.

We have that $\frac{S(\mu,\nu)-0.529}{1-0.529}=\frac{0.809-0.529}{1-0.529}=0.594$, $\frac{S(\mu,\rho)-0.529}{1-0.529}=\frac{0.846-0.529}{1-0.529}=0.673$, and $\frac{S(\nu,\rho)-0.529}{1-0.529}=\frac{0.827-0.529}{1-0.529}=0.633$.

Sub-Saharan Africa

We have that $S(\mu,\nu)=1-510/1892=0.730$, $S(\mu,\rho)=1-184/1892=0.903$, and $S(\nu,\rho)=1-553/1886=0.707$. Here we have 43 countries and $\frac{1}{2}+\frac{1}{2n}=\frac{1}{2}+\frac{1}{86}=0.512$.

We have that $\frac{S(\mu,\nu)-0.512}{1-0.512}=\frac{0.730-0.512}{1-0.512}=0.447$, $\frac{S(\mu,\rho)-0.512}{1-0.512}=\frac{0.903-0.512}{1-0.512}=0.801$, and $\frac{S(\nu,\rho)-0.512}{1-0.512}=\frac{0.707-0.512}{1-0.512}=0.400$.

Appendix 2.1 Sustainable Development Goals G_1 : *No Poverty*

Internationally comparable poverty lines above $1.90 PPP per day.

G_2 : *Zero Hunger*

Agricultural yield gaps by cropping system
Resource use efficiency (nutrients, water, energy)
Food loss and food waste
Greenhouse gas emissions from land use
Diets and nutrient deficiencies

G_3 : *Good Health and Well-Being*

Affordability of healthcare

G_4 : *Quality Education*

Internationally comparable primary and secondary education outcomes
Early childhood development

G_5 : *Gender Equality*

Gender pay gap and other empowerment measures
Violence against women

G_6 : *Clean Water and Sanitation*

Water embedded in trade adjusted for environmental impact
Quality of drinking water and surface water

G_7 : *Affordable and Clean Energy*

G_8 : *Decent Work and Economic Growth*

Decent work
Child labor

G_9 : *Industry, Innovation and Infrastructure*

G_{10} : *Reduced Inequalities*

Wealth inequality
Vertical mobility

G_{11} : *Sustainable Cities and Communities*

G_{12} : *Responsible Consumption and Production*

Environmental impact of material flows
Recycling and re-use (circular economy)
Chemicals

G_{13} : *Climate Action*
Leading indicators for decarbonization
Greenhouse gas emissions from land use
Climate vulnerability metrics
G_{14} : *Life Below Water*
Maximum sustainable yields
Impact of high sea and border fishing
Protected areas by level of protection
G_{15} : *Life on Land*
Leading indicators on ecosystem health
Trade in endangered species
Protected areas by level of protection
G_{16} : *Peace, Justice, and Strong Institutions*
Modern slavery and human trafficking
Access to justice
Financial secrecy
G_{17} : *Partnerships and Goals*
Nonconcessional development finance
Climate finance
Unfair tax competition
Development impact of trade practices

Chapter 3
Global Climate Risk

3.1 Risk Models

In [134], 67 developed, emerging, and frontier market countries were ranked for their vulnerability to climate change risks. All countries are being impacted by climate change but some face much more serious problems than others. In [134], countries are scored that are most exposed to climate change risks, as well as those best placed to respond to them.

In this chapter, we place the findings in [134] into a mathematical model based on mathematics of uncertainty. This allows one to apply mathematical techniques to the study of climate risks.

The following is from [[134], pp. 16, 21, 34, 39]. The numbers in parentheses are the weights given to the variables in [134].

Let R denote risk.

P denote Physical Risk (0.25)

S denote Sensitivity to Extreme Events (0.25)

E denote Energy Transition Risks (0.25)

C denote Potential to Respond to Climate Risks (0.25)

P_1 : Temperature 0.35

P_{11} : Abs 0.25

P_{12} : Change 0.15

P_2 : Water Availability 0.50

P_{21} : Abs 0.25

P_{22} : Change 0.25

P_3 : Extreme Events 0.15

S_1 : Cost 0.40

S_2 : Deaths 0.40

S_3 : People Effected 0.30

E_1 : Fossil Rents 0.33 (0.11 is Chg)

E_2 : Fossil in Export 0.33 (0.11 is Chg)

J. N. Mordeson and S. Mathew, *Mathematics of Uncertainty for Coping with World Challenges*, Studies in Systems, Decision and Control 353,
https://doi.org/10.1007/978-3-030-68684-0_3

E_3 : Fossil in Energy 0.33 (0.11 is Chg)
C_1 : (0.50)
C_2 : (0.50)
$R = 0.25P + 0.25S + 0.25E + 0.25C$
$P = 0.35P_1 + 0.50P_2 + 0.15P_3$
$P_1 = 0.25P_{11} + 0.15P_{12}$
$P_2 = 0.25P_{21} + 0.25P_{22}$

Three experts, E_1, E_2, E_3 ranked the importance of P, S, E, and C. Their ratings are given in the following table.

AHP	E_1	E_2	E_3	Row Avg.
P	7	7	7	7
S	7	8	6	7
E	7	9	8	8
C	7	6	5	6
Col. Sum	28	30	26	28

AHP equation:

$$R = 0.25P + 0.25S + 0.29E + 0.21C$$

Guiasu/Dempster	E_1	E_2	E_3	Row Avg.	Row Product
P	0.25	0.23	0.27	0.25	0.015525
S	0.25	0.27	0.23	0.25	0.015525
E	0.25	0.30	0.31	0.29	0.02325
C	0.25	0.20	0.19	0.21	0.0095
Col. Sum					0.0638

Guiasu equation:

$$R = 0.25P + 0.25S + 0.29E + 0.21C$$

We next show how to develop the Dempster Method from the Guiasu table. Suppose we have n experts $E_1, \ldots, E_n$ and m subfactors $F_1, \ldots, F_m$ We consider that each expert E_j has $\{\{F_1\}, \ldots, \{F_m\}\}$ as its set of focal elements. An entry $m_j(F_i)$ in the Guiasu table, $i = 1, \ldots, m$; $j = 1, \ldots, n$, can be interpreted as a basic probability assignment for expert E_j and focal point $\{F_i\}$ since $\sum_{i=1}^{m} m_j(F_i) = 1$, that is, the entries in each column the Guiasu table sum to 1. Consider the row product $m_1(F_i) \ldots m_n(F_i), i = 1, \ldots, m,$of the Guiasu table. Since $\sum_{i=1}^{m} m_j(F_i) = 1$ for $j = 1, \ldots, n$, we have that $\prod_{j=1}^{n} \sum_{i=1}^{m} m_j(F_i) = 1$. Thus $1 = m_1(F_1) \ldots m_n(F_1) + \cdots + m_1(F_m) \ldots m_n(F_m) + K$, where

$$K = \prod_{j=1}^{n} \sum_{i=1}^{m} m_j(F_i) - [m_1(F_1) \ldots m_n(F_1) + \cdots + m_1(F_m) \ldots m_n(F_m)]$$

That is, $1 - K = m_1(F_1) \ldots m_n(F_1) + \cdots + m_1(F_m) \ldots m_n(F_m)$ and hence we see that if we divide each entry in the product column of the Guiasu table by $1 - K$, we normalize each entry in the product column of the Guiasu table. Thus from Sect. 1.2 of Chap. 1, $m_{1,\ldots,n}(F_i) = m_1(F_i) \ldots m_n(F_i)$ is a basic probability assignment for the joint basic probability distribution, $i = 1, \ldots, m$.

Dempster equation:

$$R = 0.24P + 0.24S + 0.36E + 0.15C.$$

Yen	E_1	E_2	E_3	Row Avg.
P	1	0.77	0.87	0.88
S	1	0.90	0.74	0.88
R	1	1	1	1
C	1	0.67	0.61	0.76
Col.Sum	4			3.52

Yen equation:

$$R = 0.25P + 0.25S + 0.28E + 0.22C$$

3.2 Risk

All countries are being impacted by climate change but some are facing more acute challenges than others. In [134], countries were identified and scored with respect to the most exposed to climate change risks, as well as those best placed to respond to them. It is stated in [134] that climate change manifests through rising temperatures, can alter hydrological (water) cycles and exacerbates extreme weather events. In turn this means higher risks to energy, food, and water systems, populations and the global economy. Over 2030 to 2950, the World Health Organization (WHO) expects 250,000 additional deaths per year due to climate change.

Furthermore, as the world seeks to limit climate change, a combination of climate policy and disruptive cleaner technologies, which do not use fossil fuels (particularly in the power and transportation sectors), mean that the peak for fossil fuel demand may arrive in the coming years. In [134] countries that are most vulnerable to climate change—in terms of both physical impacts and the associated energy transition risks—and which are better placed to respond to these pressures are determined.

In [[134], p. 5], a score and a rank were given for 67 countries. These countries represent almost one-third of the world's nation states, but 80% of the global population and 94% of global GDP. The overall score for each country was determined by taking the average of the four scores. The data entries in the first four columns of the table in [[134], p. 5] is made more compatible by using the formula $\frac{\#-\min}{\max-\min}$, where for a given column, min is the smallest number in that column, max is the largest element in that column, and # is the number of interest in that column. For example, the entry for India in the Physical Impacts column is 3.84. The minimum value in the column is 2.08 and the maximum value is 8.22. Hence $\frac{3.82-2.08}{8.22-2.08} = 0.289$ is the entry in Physical Impacts column in the following table. The AHP equation and the Guiasu equation are the same and the Yen equation is nearly the same. Thus we use only the Guiasu and Dempster equations. The score values in the following table are determined from the scores in the table below by taking their weighted average using the Guiasu method and the Dempster method. In the table, G stands for Guiasu and D stands for Dempster.

In Table 3.1 below, we have I: Physical impacts, II: Sensitivity to extreme events, III: Energy transition risk, IV: Potential to respond to climate risks, V: Overall climate vulnerability.

Let μ denote the Guiasu ranking and ν the Dempster ranking. We have that $M(\mu, \nu) = \frac{2177}{2379} = 0.915$. Here $n = 67$ countries and $\frac{n+1}{3n-1} = \frac{67+1}{201-1} = \frac{68}{200} = 0.340$.

$S(\mu, \nu) = 1 - \frac{202}{4566} = 1 - .044 = 0.956$ countries and $\frac{1}{2} + \frac{1}{2n} = \frac{1}{2} + \frac{1}{134} = 0.507$.

Now

$$\begin{aligned}(S(\mu, \nu) - 0.507)/(1 - 0.507) &= (0.956 - 0.507)/(1 - 0.507) = 0.911 \\ &> 0.871 = (0.915 - 0.340)/(1 - 0.340) \\ &= (M(\mu, \nu) - 0.340)/(1 - 0.340).\end{aligned}$$

Let ρ denote the Risk ranking given [[19], p. 5]. We have that
$M(\mu, \rho) = \frac{2162}{2394} = 0.903$,
$S(\mu, \rho) = 1 - \frac{230}{4540} = 0.949$.
Now

$$\begin{aligned}(S(\mu, \rho) - 0.507)/(1 - 0.507) &= (0.949 - 0.507)/(1 - 0.507) = 0.955 \\ &> 0.853 = (0.903 - 0.340)/(1 - 0.340) \\ &= (M(\mu, \rho) - 0.340)/(1 - 0.340).\end{aligned}$$

We also have
$M(\nu, \rho) = \frac{2078}{2478} = 0.839$,
$S(\nu, \rho) = 1 - \frac{396}{4556} = 0.913$.
Now

$$\begin{aligned}(S(\nu, \rho) - 0.507)/(1 - 0.507) &= (0.913 - 0.507)/(1 - 0.507) = 0.824 \\ &> 0.756 = (0.839 - 0.340)/(1 - 0.340) \\ &= (M(\nu, \rho) - 0.340)/(1 - 0.340).\end{aligned}$$

Table 3.1 Climate vulnerability

Country	I	II	III	IV	V	
	Score $G \setminus D$	Score $G \setminus D$	Score $G \setminus D$	Score $G \setminus D$	Score $G \setminus D$	Rank $G \setminus D$
Weights	0.25\0.24	0.25\0.24	0.29\0.36	.21\0.15		
India	0.289	0.160	0.468	0.110	0.2711\0.2957	1\2
Pakistan	0.242	0.069	0.860	0.018	0.3309\0.3908	7\12
Philippines	0.256	0	0.811	0.122	0.3248\0.3755	4\10
Bangladesh	0.394	0 0.097	0.624	0.102	0.3251\3614	5\9
Oman	0.135	0.211	0.189	0.532	0.2530\0.2332	2\1
Sri Lanka	0.313	0.082	0.858	0.040	0.3560\0.4138	11\16
Columbia	0.657	0.220	0.227	0.134	0.3132\0.3155	3\3
Mexico	0.385	0.335	0.476	0.086	0.3361\0.3607	9\8
Kenya	0.136	0.341	0.897	0	0.3794\0.4418	12\22
S.Africa	0.282	0.493	0.386	0.122	0.3313\0.3467	8\5
Thailand	0.547	0.055	0.661	0.222	0.3888\0.4199	15\18
Israel	0.028	0.438	0.438	0.401	0.3277\0.333	6\4
Lebanon	0.201	0.698	0.378	0.010	0.3365\0.3569	10\7
Vietnam	0.493	0.119	0.680	0.207	0.3937\0.427	16\20
Nigeria	0.154	0.599	0.609	0.110	0.3880\0.4207	14\19
Morocco	0.318	0.528	0.605	0.130	0.4142\0.4448	19\23
Indonesia	0.494	0.426	0.588	0.130	0.4278\0.4565	21\25
Egypt	0.167	0.829	0.481	0.053	0.3996\0.4244	18\21
Brazil	0.644	0.389	0.609	0.058	0.4470\0.4807	25\29
Serbia	0.651	0.244	0.674	0.214	0.4642\0.4945	29\31
Malaysia	0.494	0.534	0.217	0.370	0.3976\0.3842	17\11
Peru	0.670	0.433	.461	0.169	0.4449\0.4606	27\27
Bahrain	0.410	1	0	0.450	0.447\0.41	24\15
S.Arabia	0.122	0.639	0.193	0.654	0.3836\0.3538	13\6
Greece	0.573	0.512	0.444	0.225	0.4473\0.4586	26\26
China	0.718	0.195	0.618	0.344	0.4797\0.4982	30\32
Tunesia	0.323	0.829	0.425	0.125	0.4375\0.4528	23\24
Argentina	0.664	0.517	0.640	0.173	0.5172\0.5452	35\40
Australia	0.594	0.234	0.240	0.849	0.4549\0.4166	28\17
Mauritius	0.315	0.773	0.556	0.288	0.4937\0.5096	31\34
UK	0.469	0.404	0.680	0.506	0.5217\0.5356	37\39
Poland	0.539	0.394	0.637	0.477	0.5181\0.5301	36\37
Qatar	0	1	0.169	0.604	0.4259\0.3954	20\13
Czech Rep.	0.302	0.357	0.848	0.580	0.5325\0.556	40\43
Russia	0.929	0.409	0.348	0.370	0.5131\0.5070	34\33

(continued)

Table 3.1 (continued)

Country	I	II	III	IV	V	
	Score $G \setminus D$	Score $G \setminus D$	Score $G \setminus D$	Score $G \setminus D$	Score $G \setminus D$	Rank $G \setminus D$
Weights	0.25\0.23	0.25\0.23	0.29\0.36	0.21\0.15		
Portugal	0.638	0.512	0.624	0.311	0.5338\0.5528	41\42
Kuwait	0.073	1	0.124	0.631	0.4367\0.4008	22\14
Jordan	0.192	1	0.652	0.169	0.5226\0.5517	38\41
USA	0.766	0.222	0.513	0.667	0.5358\0.5271	43\35
Belgium	0.398	0.594	0.547	0.580	0.5248\0.5273	39\36
Kazakhstan	0.690	0.696	0.212	0.465	0.5056\0.4835	33\30
Japan	0.628	0.572	0.472	0.468	0.5352\0.5335	42\38
France	0.451	0.348	0.976	0.512	0.5903\0.6262	46\49
Hungary	0.664	0.398	0.916	0.332	0.6009\0.6408	49\52
Romania	0.688	0.449	0.811	0.388	0.6009\0.6293	50\50
Slovenia	0.609	0.387	0.811	0.527	0.5949\0.6152	48\47
Italy	0.516	0.728	0.770	0.270	0.591\0.6225	47\48
Turkey	0.401	0.815	0.708	0.317	0.5759\0.6003	44\46
UAE	0.083	1	0.277	0.713	0.5008\0.4713	32\28
Croatia	0.688	0.699	0.717	0.309	0.6196\0.6438	51\53
Chile	0.883	0.301	0.886	0.521	0.6626/0.6882	54\56
Singapore	0.244	1	0.444	0.749	0.5970\0.5765	45\44
Spain	0.628	0.794	0.815	0.388	0.6733\0.6999	56\57
Germany	0.570	0.653	0.773	0.639	0.6641\0.6744	55\54
Lithuania	0.673	0.780	0.669	0.555	0.6738\0.6796	57\55
Netherlands	0.571	0.909	0.335	0.711	0.6165\0.5883	52\45
S. Korea	0.523	0.747	0.588	0.782	0.6522/0.6402	53\51
Austria	0.596	0.694	0.919	0.657	0.7270\0.7465	58\59
Canada	1	0.557	0.652	0.71‘6	0.7219\0.7230	59\58
Switzerland	0.518	0.806	1	0.687	0.7653\0.7887	60\62
Denmark	0.407	0.980	0.888	0.778	0.7676\0.7770	61\61
Ireland	0.766	0.872	0.787	0.760	0.7973\0.7964	62\63
New Zealand	0.813	0.653	0.925	0.885	0.8206\8358	64\64
Estonia	0.841	0.978	0.929	0.580	0.8460\0.8667	65\67
Norway	0.821	0.940	0.532	1	0.8035\0.7719	63\60
Sweden	0.831	0.984	0.850	0.746	0.8569\0.8621	67\66
Finland	0.883	1	0.768	0.757	0.8524\0.8505	66\65

3.3 Global Climate Risk Index

The German watch Global Climate Risk Index (CRI) is an analysis based on one of the most reliable data sets available on the impacts of extreme weather events and associated-economic data, [48]. It is stated in [48] that the CRI indicates a level of exposure and vulnerability to extreme events, which countries should understand as warnings in order to be prepared for more frequent and/or more severe events in the future.

In this section, we compare the Climate Risk Index in [48] with the Risk ranking in [134]. We use two fuzzy similarity relations to measure their similarity.

The CRI is determined from the analysis of four indicators stated immediately below, [[48], p. 24].

R : Risk

R_1 : Fatalities (annual average)

R_2 : Fatalities per 100.000 inhabitants (annual average)

R_3 : Losses in US$ million (PPP) (annual average)

R_4 : Losses per unit GDP (annual average)

The weights of the four indicators R_1, R_2, R_3, R_4 are given in [[20], p. 22] to be $1/6, 1/3/1/6, 1/3,$ respectively

$R = \frac{1}{6}R_1 + \frac{1}{3}R_2 + \frac{1}{6}R_3 + \frac{1}{3}R_4$ page 22

The equation for R is used to determine the CRI scores given in [[48], Table 5]. CRI in the table below is from page [[48], p. 34] and Risk is from [[134], p. 21].

Table 3.2 Risk and CRI rankings

Country	Risk	CRI
Philippines	1	5
Thailand	2	13
Pakistan	3	8
Sri Lanka	4	31
Bangladesh	5	7
Vietnam	6	9
India	7	14
China	8	37
Oman	9	29
Columbia	10	49
USA	11	27
Australia	12	36
Serbia	13	73
Chile	14	94

(continued)

Table 3.2 (continued)

Country	Risk	CRI
Mexico	15	53
Kenya	16	45
France	17	18
Czech Rep.	18	85
Slovenia	19	43
Brazil	20	90
Poland	21	72
Hungary	22	64
UK	23	60
Russia	24	33
Indonesia	25	69
Peru	26	45
Isreal	27	138
Romania	28	37
South Africa	29	78
Greece	30	87
Portugal	31	22
Argentina	32	83
Morocco	33	53
Malaysia	34	116
Canada	35	100
Japan	36	93
Belgium	37	60
Nigeria	38	127
S.Arabia	39	115
Germany	40.5	25
New Zealand	40.5	88
Austria	42	51
Kazakhstan	43	157
Lebanon	44	141
Croatia	45	35
Italy	46	28
S. Korea	47	80
Mauritius	48	118
Lithuania	49	117
Spain	50	34
Switzerland	51	41
Turkey	52	121
Egypt	53	156

(continued)

Table 3.2 (continued)

Country	Risk	CRI
Tunesia	53	136
Ireland	55	125
Netherlands	56	71
Norway	57	151
Estonia	58	161
Denmark	59	131
Sweden	60	146
Qatar	64	181
Bahrain	64	140
Kuwait	64	179
UAE	64	169
Jordan	64	142
Singapore	64	180
Finland	64	167

Table 3.3 Risk and CRI rankings

Country	Risk	CRI
Philippines	1	1
Thailand	2	5
Pakistan	3	3
Sri Lanka	4	13
Bangladesh	5	2
Vietnam	6	4
India	7	6
China	8	18.5
Oman	9	12
Columbia	10	24
USA	11	10
Australia	12	17
Serbia	13	34
Chile	14	43.5
Mexico	15	26.5
Kenya	16	22.5
France	17	7
Czech Rep.	18	38

(continued)

Table 3.3 (continued)

Country	Risk	CRI
Slovenia	19	21
Brazil	20	41
Poland	21	33
Hungary	22	30
UK	23	28.5
Russia	24	14
Indonesia	25	31
Peru	26	22.5
Isreal	27	53.5
Romania	28	18.5
South Africa	29	35
Greece	30	39
Portugal	31	8
Argentina	32	37
Morocco	33	26.5
Malaysia	34	46
Canada	35	43.5
Japan	36	42
Belgium	37	28.5
Nigeria	38	51
S.Arabia	39	45
Germany	40.5	9
New Zealand	40.5	40
Austria	42	25
Kazakhstan	43	61
Lebanon	44	56
Croatia	45	16
Italy	46	11
S. Korea	47	36
Mauritius	48	48
Lithuania	49	47
Spain	50	15
Switzerland	51	20
Turkey	52	49
Egypt	53	60
Tunesia	53	53.5
Ireland	55	50

(continued)

Table 3.3 (continued)

Country	Risk	CRI
Netherlands	56	32
Norway	57	59
Estonia	58	62
Denmark	59	52
Sweden	60	58
Qatar	64	67
Bahrain	64	55
Kuwait	64	65
UAE	64	64
Jordan	64	57
Singapore	64	66
Finland	64	63

We have $M(\mu, \nu) = 0.748$. We have 67 countries and $\frac{n+1}{3n-1} = \frac{67+1}{201-1} = \frac{68}{200} = 0.340$.

We have $S(\mu, \nu) = 0.857$. We have 67 countries and $\frac{1}{2} + \frac{1}{2n} = \frac{1}{2} + \frac{1}{134} = 0.507$. Now

$$\begin{aligned}(S(\mu, \nu) - 0.507)/(1 - 0.507) &= (0.857 - 0.507)/(1 - 0.507) = 0.710 \\ &> 0.618 = (0.748 - 0.340)/(1 - 0.340) \\ &= (M(\mu, \nu) - 0.340)/(1 - 0.349).\end{aligned}$$

Thus $S(\mu, \nu)$ gives a relatively higher similarity measure than $M(\mu, \nu)$.

Chapter 4
Human Trafficking and Modern Slavery

4.1 Human Trafficking

Trafficking in persons disproportionately affects women and girls; 71% of all victims detected worldwide are female. The 2016 UNODC Global report on Trafficking estimates that 51% of all detected trafficking victims are women and 20% are girls.

Out of the 17 SDGs, trafficking in persons is specifically mentioned in three targets under three goals G_5 Gender Equality, G_8 Decent Work and Economic Growth, and G_{16} Peace, Justice, and Strong Institutions. However many other SDG targets and goals are relevant to addressing trafficking in persons. This issue is rooted in development issues at-large including poverty, education, child labor, abuse and exploitation, gender equality and discrimination, migration and the effects of climate change, [68]. Other SDGs mentioned in [68] that contribute to combatting trafficking in persons are G_{10}, G_4, and G_{17}. In [144], it is also stated that G_{12} is also important in combatting human trafficking.

In [133], Stakeholder Forum created a methodology to enable scores to be assigned to each of the different targets and goals according to their significance in different contexts. The method has been discussed in Chap. 1. We also use the analytic hierarchy process, the Guiasu method, and the Yen method to determine which countries are, in the opinion of the experts, the best in achieving sustainability with respect to human trafficking. The equations involve the SDGs that are pertinent to human trafficking.

AHP equation:

$$G = 0.06G_1 + 0.06G_4 + 0.16G_5 + 0.07G_6 + 0.17G_8 \\ +0.06G_{10} + 0.07G_{11} + 0.06G_{12} + 0.24G_{16} + 0.06G_{17}.$$

Guiasu equation:

$$G' = 0.06G_1 + 0.06G_4 + 0.14G_5 + 0.05G_6 + 0.20G_8 \\ +0.04G_{10} + 0.05G_{11} + 0.04G_{12} + 0.030G_{16} + 0.04G_{17}.$$

J. N. Mordeson and S. Mathew, *Mathematics of Uncertainty for Coping with World Challenges*, Studies in Systems, Decision and Control 353,
https://doi.org/10.1007/978-3-030-68684-0_4

Yen equation:

$$G'' = 0.06G_1 + 0.05G_4 + 0.14G_5 + 0.07G_6 + 0.17G_8 + 0.06G_{10} + 0.07G_{11} + 0.06G_{12} + 0.25G_{16} + 0.06G_{17}.$$

Stakeholder equation:

$$SDG = 0.06G_1 + 0.09G_4 + 0.08G_5 + 0.09G_6 + 0.09G_8 + 0.13G_{10} + 0.09G_{11} + 0.22G_{12} + 0.09G_{16} + 0.06G_{17}.$$

We next consider the ranking of countries with respect achieving the goal of combatting human trafficking. We have no ranking for the SD Report. Thus we compare only the AHP-Guiasu-Yen ranking and the Stakeholder ranking. Due to missing data, we rerank. The following rankings are from [[113], pp. 121–127] (Tables 4.1, 4.2, 4.3, 4.4, 4.5 and 4.6).

The Scandinavian countries rank the highest in achieving the SDGs pertinent to modern slavery. In particular, Denmark and Finland ranked first and second, respectively.

Security threats arise from not only military disasters, but also from climate change, natural disasters, infectious diseases, and transnational crimes. Among those nontraditional security threats, human trafficking looms large since natural disasters and military lead to displaced people and refugees, who are particularly vulnerable, [14]. Singapore ranks the highest in the region with respect to achieving the SDGs pertinent to human trafficking.

Data concerning trafficking in this region can be found in [61]. We find that Belarus and Moldova rank high in achieving the SDGs pertinent to human trafficking.

South America is one of the biggest source and destination locations in the world. Currently, Latin America is the third largest source of trafficked humans. Our ranking has Cuba as the leader in the region in achieving the SDGs pertinent to human trafficking. The Government of Cuba does not fully meet the minimum standards for the elimination of trafficking. However, it is making significant steps to do so, [164].

Inaccurate information in the region can make determining facts and information difficult. Some laws and cultures in the region do not recognize certain forms of human trafficking as trafficking, [47]. Our rankings show that Algeria is the best at achieving the SDGs pertinent to human trafficking.

Our rankings don't agree on one country that is doing the best at achieving the SDGs pertinent to human trafficking. The following is from [139]. There is a distinct trafficking pattern for different African regions. Victims from sub-Saharan Africa were found in over 60 countries within and outside Africa. According to the UN, this makes sub-Saharan Africa, 'a relevant origin for detected cases of trafficking globally.'

Table 4.1 OECD combined ranks

OECD

Country	AHP, Guiasu, Yen rank	Stakeholder rank
Australia	17.5	21.5
Austria	7	11
Belgium	11.5	13
Canada	9	9
Chile	30	31
Czech Rep.	15	6.5
Denmark	1	1
Estonia	13	18
Finland	2	2
France	17.5	13
Germany	10	10
Greece	35	36
Hungary	29	17
Iceland	3.5	6.5
Ireland	16	15.5
Israel	33	33
Italy	31	28
Japan	19	19
Korea Rep.	28	26
Latvia	25	29
Lithuania	27	30
Luxembourg	22	25
Mexico	36	35
Netherlands	11.5	8
New Zealand	5	13
Norway	6	3
Poland	24	27
Portugal	20	24
Slovak Rep.	26	15.5
Slovenia	8	4.5
Spain	23	23
Sweden	3.5	4.5
Switzerland	21	20
Turkey	34	32
United Kingdom	14	21.5
United States	32	34

Table 4.2 East and South Africa combined ranks

East and South Asia		
Country	AHP, Guiasu, Yen rank	Stakeholder rank
Bangladesh	18	8
Bhutan	8	7
Brunei Darussiam	6.5	1
Cambodia	15	14
China	2	6
India	17	9.5
Indonesia	11	16
Korea, Dem. Rep.		
Lao PDR	13	12
Malaysia	6.5	9.5
Maldives	3.5	2
Mongolia	9	17
Myanmar	12	13
Nepal	16	5
Pakistan	19	19
Philippines	14	18
Singapore	1	3
Sri Lanka	10	11
Thailand	3.5	15
Timor Leste		
Vietnam	5	4

Table 4.3 Eastern Europe and Central Asia combined ranks

Eastern Europe and Central Asia		
Country	AHP, Guiasu, Yen rank	Stakeholder rank
Afghanistan	23	23
Albania	20.5	13
Andorra		
Armenia	20.5	15.5
Azerbaijan	17	10.5
Belarus	1	3
Boznia and Herzegovina	15.5	2
Bulgaria	9.5	20
Croatia	3	10.5
Cypris	7.5	22
Georgia	7.5	15.5

(continued)

Table 4.3 (continued)

Eastern Europe and Central Asia		
Country	AHP, Guiasu, Yen rank	Stakeholder rank
Kazakhstan	9.5	21
Kyrgyz Rep.	15.5	5.5
Liecherstan		
Malta	2	5.5
Moldova	4	1
Monaco		
Montenegro	18	19
North Macedonia	19	14
Romania	5	17
Russian Federation	20	18
San Marino		
Serbia	6	8
Tajikistan	14	9
Turkmenistan	11	7
Ukraine	12.5	4
Uzbekistan	12.5	12

Table 4.4 Latin America and the Caribbean combined ranks

Latin America and the Caribbean		
Country	AHP, Guiasu, Yen rank	Stakeholder rank
Antigua and Barbuda		
Argentina	5	4
Bahamas, The Barbados		
Belize	18	16
Bolivia	15	12
Brazil	16	20
Columbia	13.5	19
Costa Rica	2	6
Cuba	1	1
Dominica		
Dominican Rep.	11	7
Ecuador	4	2
El Salvador	12	8
Grenada		
Guatemala	21	22.5

(continued)

Table 4.4 (continued)

Latin America and the Caribbean		
Country	AHP, Guiasu, Yen rank	Stakeholder rank
Guyana	19	15
Haiti	23	22.5
Honduras	20	19
Jamaica	6	5
Nicaragua	9	13
Panama	13.5	14
Paraguay	17	17.5
Peru	10	17.5
St. Kitts and Nevis		
St. Lucia		
St. Vincent and the Grenadines		
Suriname	8	11
Trinidad and Tobago	7	3
Uruguay	3	10
Venezuela, RB	22	21

Table 4.5 Middle East and North Africa combined ranks

Middle East and North Africa		
Country	AHP, Guiasu, Yen rank	Stakeholder rank
Algeria	1	1
Bahrain	6	3
Egypt, Arab Rep.	11	13
Iran, Islamic Rep.	9.5	11
Iraq	13	2
Jordan	7.5	9.5
Kuwait	14	14
Lebanon	7.5	9.5
Libya		
Morocco	5	6
Oman	9.5	8
Qatar	4	7
Saudi Arabia	12	12
Syrian Arab Rep.	15	15
Tunisia	3	5
United Arab Emirates	2	4
Yemen Rep.	16	16

Table 4.6 Sub-Saharan Africa combined ranks continued

Sub-Saharan Africa		
Country	AHP, Guiasu, Yen rank	Stakeholder rank
Angola	36	41
Benin	35	37
Botswana	5	13
Burkina Faso	25	24
Burundi	19	6
Cabo Verde	2	2
Cameroon	22	36
Central African Rep.	43	35
Chad	42	23
Comoros	17	34
Congo Democratic Rep.	41	43
Congo Rep.	29	31
Cote d'lvoire	23	26
Djibouti	13	30
Equatorial Guinea		
Eritrea		
Eswatini	18	14
Ethiopia	20	4
Gabon	12	28
The Gambia	16	33
Ghana	6	27
Guinea	32	20
Guinea-Bissau		
Kenya	11	32
Lesotho	24	11
Liberia	39	5
Madagascar	30	29
Malawi	8	7.5
Mali	34	7.5
Mauritania	37	12
Mauritius	3	10
Mozambique	26	19
Namibia	1	18
Niger	31	9
Nigeria	40	17
Rwanda	10	22
Sao Tome and Principe	4	3
Senegal	15	39

(continued)

Table 4.6 (continued)

Sub-Saharan Africa		
Country	AHP, Guiasu, Yen rank	Stakeholder rank
Seychelles		
Sierra Leone	33	25
Somalia		
South Africa	7	16
South Sudan		
Sudan	38	21
Tanzania	14	15
Togo	28	40
Uganda	21	38
Zambia	27	42
Zimbabwe	9	1

Similarity Measures Once again let μ denote the ranking for the AHP, Guiasu, Yen method and ν the ranking for the stakeholder. We provide the similarity measure of μ and ν for the various regions.

OECD

We find that $M(\mu, \nu) = \frac{611}{721} = 0.847$. The smallest possible similarity measure for 36 countries with no ties in the ranking can be obtained by substituting 36 for n in the formula $\frac{n+2}{3n+2}$. One obtains $38/110 = 0.345$.

We also have that $S(\mu, \nu) = 1 - \frac{110}{1332} = 0.917$. The smallest possible similarity measure for 36 countries with no ties can be substituting 36 for n in the formula $\frac{n/2+1}{n+1}$. One obtains $\frac{19}{37} = 0.514$.

We also have that $\frac{M(\mu,\nu)-0.345}{1-0.345} = \frac{0.847-0.345}{1-0.345} = \frac{0.502}{0.655} = 0.766$ and $\frac{S(\mu,\nu))-0.514}{1-0.514} = \frac{0.917-0.514}{1-0.514} = \frac{0.403}{0.486} = 0.829$.

East and South Asia

We have that $M(\mu, \nu) = \frac{156}{224} = 0.696$. Here we have 19 countries and $\frac{n+1}{3n-1} = \frac{19+1}{57-1} = \frac{20}{56} = 0.357$.

We have that $S(\mu, \nu) = 1 - \frac{79}{380} = 0.792$. Here we have 19 countries and the smallest number $S(\mu, \nu)$ can be is found by substituting 19 into the formula $\frac{1}{2} + \frac{1}{2n}$ for n. We obtain $\frac{1}{2} + \frac{1}{38} = 0.526$.

We also have that $\frac{M(\mu,\nu)-0.357}{1-0.357} = \frac{0.696-0.357}{1-0.357} = \frac{0.339}{0.643} = 0.527$ and $\frac{S(\mu,\nu))-0.526}{1-0.526} = \frac{0.792-0.526}{1-0.526} = \frac{0.266}{0.474} = 0.561$.

Eastern Europe and Central Asia

We have that $M(\mu, \nu) = \frac{203.5}{348.5} = 0.584$. Here we have 23 countries and $\frac{n+1}{3n-1} = \frac{23+1}{69-1} = \frac{24}{68} = 0.353$.

We have that $S(\mu, \nu) = 1 - \frac{143}{550} = 0.740$. Here we have $\frac{1}{2} + \frac{1}{2n} = \frac{1}{2} + \frac{1}{46} = 0.522$.

We also have that $\frac{M(\mu,\nu)-0.353}{1-0.353} = \frac{0.584-0.353}{1-0.353} = \frac{0.231}{0.647} = 0.357$ and $\frac{S(\mu.\nu)-0.522}{1-0.522} = \frac{0.740-0.522}{1-0.522} = \frac{0.218}{0.478} = 0.456$.

Latin America and the Caribbean

We have that $M(\mu, \nu) = \frac{244.5}{307.5} = 0.652$. Here we have 23 countries and $\frac{n+1}{3n-1} = \frac{23+1}{69-1} = \frac{24}{68} = 0.353$.

We have that $S(\mu, \nu) = 1 - \frac{64}{542} = 0.882$. Here we have $\frac{1}{2} + \frac{1}{2n} = \frac{1}{2} + \frac{1}{46} = 0.522$.

We also have that $\frac{M(\mu,\nu)-0.353}{1-0.353} = \frac{0.652-0.353}{1-0.353} = \frac{0.299}{0.647} = 0.462$ and $\frac{S(\mu.\nu))-0.522}{1-0.522} = \frac{0.882-0.522}{1-0.522} = \frac{0.360}{0.478} = 0.753$.

Middle East and North Africa

We have that $M(\mu, \nu) = \frac{120.5}{151.5} = 0.795$. Here we have 16 countries and $\frac{n+2}{sn+2} = \frac{16+2}{48+2} = \frac{18}{50} = 0.360$.

We have that $S(\mu, \nu) = 1 - \frac{31}{272} = 0.886$. We have $\frac{n/2+1}{n+1}\frac{8+1}{16+1} = 0.529$.

We also have that $\frac{M(\mu,\nu)-0.360}{1-0.360} = \frac{0.795-0.360}{1-0.360} = \frac{0.435}{0.640} = 0.680$ and $\frac{S(\mu.\nu))-0.529}{1-0.529} = \frac{0.886-0.529}{1-0.529} = \frac{0.357}{0.471} = 0.758$.

Sub-Saharan Africa

We have that $M(\mu, \nu) = \frac{687}{1205} = 0.570$. Here we have 43 countries and $\frac{n+1}{3n-1} = \frac{43+1}{129-1} = \frac{44}{128} = 0.344$.

We have that $S(\mu, \nu) = 1 - \frac{518}{1892} = 0.726$. We have $\frac{1}{2} + \frac{1}{2n} = \frac{1}{2} + \frac{1}{86} = 0.512$.

We also have that $\frac{M(\mu,\nu)-0.344}{1-0.344} = \frac{0.570-0.344}{1-0.344} = \frac{0.226}{0.656} = 0.345$ and $\frac{S(\mu.\nu))-0.512}{1-0.512} = \frac{0.726-0.512}{1-0.512} = \frac{0.214}{0.488} = 0.439$.

4.2 Modern Slavery

It is stated by the Intergovernmental Panel on Climate Change that in the future millions of people will be forced to relocate due to effects on climate change, including shoreline erosion, coastal flooding or disruption of normal farming practices. It is predicted that this crisis in making will affect 150–200 million men, women, and children by 2050, or roughly one in 45 persons. Displacement leads to a considerable increase in human trafficking and modern slavery. The U. N. Environment Program has indicated that trafficking may increase by 20–30% during disasters, [55].

The poorest countries, those least responsible for greenhouse emissions causing climate change, that are bearing the greatest burden and are the first to deal with forced migration. In most poorest countries, those living in rural areas who depend on climate-sensitive resources such as water supplies and farming for their basic livelihoods are particularly vulnerable to climate change (Tables 4.7, 4.8 and 4.9).

If an environmental disaster occurs, people are forced to flee without legal authorization or documents. They may be unfamiliar with the culture of the refugee community and so are extremely vulnerable to human trafficking and modern slavery. Environmental refugees are at particular risk because current international refugee agreements are such that people displaced by environmental disasters do not qualify as refugees and thus do not qualify for international aid or protection. It is virtually impossible to estimate the number of people displaced by environmental disasters who become victims of trafficking. However, data suggests that human trafficking increases by around 20–30% during disasters.

Two of the many factors that drive environmental migration are climate processes and climate events. Climate processes take place over a period of time and include desertification of land, sea-level rise, glacier melting. and growing water scarcity. Climate events, on the other hand, are sudden and dramatic occurrences such as floods, storms, hurricanes and typhoons, which force people to leave their land quickly.

An increase in the frequency and severity of floods and storms, drought, and the resulting land degradation will force migration of the world's most vulnerable people. Greenhouse gas emissions are making our oceans more acidic and are destroying coral reefs, affecting communities' access to fish and other food. Rising temperatures cause glaciers to shrink and contribute to sea-level rise and consequently cause people to move from their homes. Intense heat waves and droughts are drastically impacting the livelihoods of farmers who depend on agriculture for their survival.

In [55, 165], it was predicted that there could be as many as 250 million people refugees by 2050. In [55, 165], it was warned that by the end of the century climate change would force one-eighth of the world's population, as many as 1.54 billion people, largely from the tropics to migrate more than 620 miles from their current homes. Some of these climate migrants, finding themselves desperate for security and work, could become victims of human trafficking, [55, 165].

Sustainable Development Goal 8 is Decent Work and Economic Growth. Target 8.7 states 'take immediate action and effective measures to eradicate forced labor, and modern slavery and human trafficking and secure the prohibition and elimination of the worst forms of child labor, including recruitment and use of child soldiers, and by 2025 end forced labor in all forms'.

We have no ranking for the SD Report. Thus we compare only the AHP-Guiasu-Yen ranking and the Stakeholder ranking. Due to missing data, we rerank.

The following equations were determined in [113] and the independent variables were based in large part on [152].

AHP equation:

$$G = 0.08G_1 + 0.08G_4 + 0.16G_5 + 0.26G_8 + 0.14G_{10} + 0.29G_{16}.$$

Guiasu equation:

$$G' = 0.06G_1 + 0.05G_4 + 0.11G_5 + 0.32G_8 + 0.09G_{10} + 0.037G_{16}.$$

Yen equation:

$$G'' = 0.08G_1 + 0.07G_4 + 0.14G_5 + 0.28G_8 + 0.12G_{10} + 0.32G_{16}.$$

Stakeholder equation:

$$SDG = 0.13G_1 + 0.09G_4 + 0.14G_5 + 0.24G_8 + 0.08G_{10} + 0.32G_{16}.$$

The rankings in the following tables are from [[15], pp. 161–165] (see Tables 4.7, 4.8, 4.9, 4.10, 4.11 and 4.12).

The Scandinavian countries rank the highest in achieving the SDGs pertinent to modern slavery. A recent report provides new estimates of child labor and human trafficking in global supply lines. The report indicates that a significant share of child labor and human trafficking in global supply chains at their lower tiers, in activities such as raw material and agriculture, [132].

Human trafficking and forced labor are acute problems in the fisheries industry, but they also affect other industries in Southeast Asia. Industries such as agriculture, manufacturing, construction, and domestic work are affected by trafficking and forced labor, [82]. Our ranking show that Singapore is the best in the region in achieving the SDGs pertinent to modern slavery.

The global estimates indicate that 4.1 per cent of all children in the Europe and Central Asia region are involved in child labor, the second lowest rate in the world. Almost all of these children are in hazardous work, the second highest prevalence rate in the world in hazardous work, only following Africa, [1]. Our ranking shows that Malta ranks the highest in the region in achieving the SDGs pertinent to modern slavery.

Cuba, Costa Rica, Trinidad and Tobago, and Uruguay rank the highest in achieving the SDGs pertinent to modern slavery. A report, [53], highlights how trafficking networks and other criminal groups profits from exploiting marginal populations, especially women and children. By some estimates, 70% of all modern slavery cases in Mexico involve organized crime groups. This includes kidnapping of girls and women for forced prostitution and mostly boys and men for forced labor.

United Arab Emirates, Bahrain, and Qatar ranked the highest in the region in achieving the SDGs pertinent to modern slavery. The following is from [77]. The Middle East and North Africa region is a volatile breeding ground for modern slavery. With government policies that promote the kafala system exploiting workers' rights and allow forced marriages to become the cultural norm, it is important to look at how they are holding the region back from achieving its development potential.

Mauritius and Sao Tome and Principe rank the highest in the region with respect to achieving the SDGs pertinent to modern slavery. However, according to [37], Africa recently recorded the highest rate of modern-day slavery. Armed conflict, state-sponsored forced labor, and forced marriages were the main causes behind the estimated 9.2 million Africans who live in servitude without choice to do so, according to the 2018 Global Slavery Index. Despite these practices being widespread, slavery has remained a largely invisible issue, in part, because it proportionately affects the most marginalized members of society, such as minorities, women, and children, [37].

Similarity Measures

Once again let μ denote the ranking for the AHP, Guiasu, Yen method and ν the ranking for the stakeholder. We provide the similarity measure of μ and ν for the various regions (Tables 4.10, 4.11 and 4.12).

OECD

We find that $M(\mu, \nu) = \frac{616}{716} = 0.860$. The smallest possible similarity measure for 36 countries with no ties in the ranking can be obtained by substituting 36 for n in the formula $\frac{n+2}{3n+2}$. One obtains $38/110 = 0.345$.

We have that $S(\mu, \nu) = 1 - \frac{102}{1332} = 0.923$. The smallest possible similarity measure for 36 countries with no ties in the ranking can be obtained by substituting 36 for n in the formula $\frac{n/2+1}{n+1}$. One obtains 0.514.

We also have that $\frac{M(\mu,\nu)-0.345}{1-0.345} = \frac{0.860-0.345}{1-0.345} = \frac{0.515}{0.655} = 0.782$ and $\frac{S(\mu.\nu))-0.514}{1-0.514} = \frac{0.923-0.514}{1-0.514} = \frac{0.409}{0.486} = 0.842$.

East and South Asia

We have that $M(\mu, \nu) = \frac{168.5}{211.5} = 0.797$. Here we have 19 countries and $\frac{n+1}{3n-1} = \frac{19+1}{57-1} = \frac{20}{56} = 0.357$.

We have that $S(\mu, \nu) = 1 - \frac{43}{380} = 0.887$. The smallest similarity measure for 19 countries with no ties in the ranking can be obtained by substituting 19 for n in the formula $\frac{1}{2} + \frac{1}{2n}$. One obtains $\frac{1}{2} + \frac{1}{38} = 0.526$.

We also have that $\frac{M(\mu,\nu)-0.357}{1-0.357} = \frac{0.797-0.357}{1-0.357} = \frac{0.440}{0.643} = 0.684$ and $\frac{S(\mu.\nu))-0.526}{1-0.526} = \frac{0.887-0.526}{1-0.526} = \frac{0.361}{0.474} = 0.762$.

Eastern Europe and Central Asia

We have that $M(\mu, \nu) = \frac{247.5}{304.5} = 0.813$. Here we have 23 countries and $\frac{n+1}{3n-1} = \frac{23+1}{69-1} = \frac{24}{68} = 0.353$.

We have that $S(\mu, \nu) = 1 - \frac{57}{552} = 0.897$. Here we have 23 countries and $\frac{1}{2} + \frac{1}{2n} = \frac{1}{2} + \frac{1}{46} = 0.522$.

We also have that $\frac{M(\mu,\nu)-0.353}{1-0.353} = \frac{0.813-0.353}{1-0.353} = \frac{0.460}{0.647} = 0.711$ and $\frac{S(\mu.\nu))-0.522}{1-0.522} = \frac{0.897-0.522}{1-0.522} = \frac{0.375}{0.478} = 0.785$.

Latin America and the Caribbean

We have that $M(\mu, \nu) = \frac{250}{302} = 0.781$. Here we have 23 countries and $\frac{n+1}{3n-1} = \frac{23+1}{69-1} = \frac{24}{68} = 0.353$.

We have that $S(\nu, \mu) = 1 - \frac{52}{553} = 0.906$. We have 23 countries and $\frac{1}{2} + \frac{1}{2n} = \frac{1}{2} + \frac{1}{46} = 0.522$.

We also have that $\frac{M(\mu,\nu)-0.353}{1-0.353} = \frac{0.781-0.353}{1-0.353} = \frac{0.428}{0.647} = 0.662$ and $\frac{S(\mu.\nu))-0.522}{1-0.522} = \frac{0.906-0.522}{1-0.522} = \frac{0.384}{0.486} = 0.790$.

Middle East and North Africa

We have that $M(\mu, \nu) = \frac{125.5}{146.5} = 0857$. Here we have 16 countries and $\frac{n+2}{3n+2} = \frac{16+2}{48+2} = \frac{18}{50} = 0.360$.

We have that $S(\mu, \nu) = 1 - \frac{21}{272} = 0.923$. We have 16 countries and $\frac{n/2+1}{n+1} = \frac{8+1}{17} = 0.529$.

We also have that $\frac{M(\mu,\nu)-0.360}{1-0.360} = \frac{0.857-0.360}{1-0.360} = \frac{0.497}{0.640} = 0.777$ and $\frac{S(\mu.\nu))-0.529}{1-0.529} = \frac{0.923-0.529}{1-0.529} = \frac{0.394}{0.471} = 0.837$.

Sub-Saharan Africa

We have that $M(\mu, \nu) = \frac{795}{1097} = 0.725$. Here we have 43 countries and $\frac{n+1}{sn-1} = \frac{43+1}{129-1} = \frac{44}{128} = 0.344$.

We have that $S(\mu, \nu) = 1 - \frac{301}{1893} = 0.841$. We have 43 countries and $\frac{1}{2} + \frac{1}{2n} = \frac{1}{2} + \frac{1}{86} = 0.512$.

We also have that $\frac{M(\mu,\nu)-0.344}{1-0.344} = \frac{0.725-0.344}{1-0.343} = \frac{0.381}{0.657} = 0.580$ and $\frac{S(\mu.\nu))-0.51}{1-0.512} = \frac{0.841-0.512}{1-0.512} = \frac{0.329}{0.488} = 0.674$.

Table 4.7 OECD combined ranks continued

OECD		
Country	AHP, Guiasu, Yen average/rank	Stakeholder average/rank
Australia	18	22.5
Austria	7.5	5
Belgium	10	11
Canada	11	6
Chile	32	32
Czech Rep.	13	8.5
Denmark	3	1
Estonia	15	22.5
Finland	1	3
France	21.5	17
Germany	17	14
Greece	34	33
Hungary	29.5	27
Iceland	2	2
Ireland	7.5	8.5
Israel	29.5	34
Italy	28	28
Japan	14	16
Korea Rep.	23	19
Latvia	24.5	30.5
Lithuania	33	29
Luxembourg	20	21
Mexico	36	36
Netherlands	12	13
New Zealand	4	15
Norway	9	7
Poland	26.5	24
Portugal	21.5	12
Slovak Rep.	24.5	25
Slovenia	6	4
Spain	26.5	26
Sweden	5	10
Switzerland	19	20
Turkey	35	35
United Kingdom	16	18
United States	31	30.5

Table 4.8 East and South Africa combined ranks

East and South Asia		
Country	AHP, Guiasu, Yen average/rank	Stakeholder average/rank
Bangladesh	18	13
Bhutan	7	9
Brunei Darussiam	9	10
Cambodia	16.5	13
China	2	2
India	15	13
Indonesia	12	11
Korea, Dem. Rep.		
Lao PDR	13	18
Malaysia	8	3
Maldives	4.5	4
Mongolia	4.5	8
Myanmar	11	17
Nepal	14	15
Pakistan	19	19
Philippines	16.5	16
Singapore	1	1
Sri Lanka	10	6
Thailand	3	5
Timor Leste		
Vietnam	6	7

Table 4.9 Eastern Europe and Central Asia combined ranks

Eastern Europe and Central Asia		
Country	AHP, Guiasu, Yen average/rank	Stakeholder average/rank
Afghanistan	23	23
Albania	22	21
Andorra		
Armenia	19	20
Azerbaijan	14	14
Belarus	3	3
Boznia and Herzegovina	17.5	11
Bulgaria	8	13
Croatia	5	4
Cypris	2	2
Georgia	12	9.5
Kazakhstan	4	12
Kyrgyz Rep.	17.5	15.5
Liecherstan		
Malta	1	1
Moldova	6	6
Monaco		
Montenegro	16	15.5
North Macedonia	20	22
Romania	9	9.5
Russian Federation	21	18
San Marino		
Serbia	7	7
Tajikistan	13	19
Turkmenistan	15	8
Ukraine	10	5
Uzbekistan	11	17

Table 4.10 Latin America and the Caribbean combined ranks

Latin America and the Caribbean		
Country	AHP, Guiasu, Yen average/rank	Stakeholder average/rank
Antigua and Barbuda		
Argentina	7	10
Bahamas, The Barbados		
Belize	18	14
Bolivia	13	16
Brazil	17	19.5
Columbia	16	18
Costa Rica	2	4
Cuba	1	1
Dominica		
Dominican Rep.	12	5
Ecuador	8	6.5
El Salvador	14	9
Grenada		
Guatemala	20	19.5
Guyana	19	17
Haiti	23	23
Honduras	21	22
Jamaica	6	6.5
Nicaragua	9	11
Panama	11	8
Paraguay	15	15
Peru	10	12
St. Kitts and Nevis		
St. Lucia		
St. Vincent and the Grenadines		
Suriname	5	13
Trinidad and Tobago	4	2
Uruguay	3	3
Venezuela, RB	22	21

Table 4.11 Middle East and North Africa combined ranks

Middle East and North Africa		
Country	AHP, Guiasu, Yen average/rank	Stakeholder average/rank
Algeria	3	3.5
Bahrain	4	1
Egypt, Arab Rep.	14	14
Iran, Islamic Rep.	9	9
Iraq	13	12
Jordan	12	13
Kuwait	11	8
Lebanon	5	7
Libya		
Morocco	10	11
Oman	6	6
Qatar	2	2
Saudi Arabia	8	5
Syrian Arab Rep.	15	15.5
Tunisia	7	10
United Arab Emirates	1	3.5
Yemen Rep.	16	15.5

Table 4.12 Sub-Saharan Africa combined ranks

Sub-Saharan Africa		
Country	AHP, Guiasu, Yen average/rank	Stakeholder average/rank
Angola	31	39
Benin	32	31
Botswana	5	4
Burkina Faso	18	3
Burundi	25	24
Cabo Verde	4	8.5
Cameroon	20	32
Central African Rep.	43	36
Chad	42	42
Comoros	25	34
Congo Democratic Rep.	40	43
Congo Rep.	38	28
Cote d'Ivoire	20	25

(continued)

Table 4.12 (continued)

Sub-Saharan Africa

Country	AHP, Guiasu, Yen average/rank	Stakeholder average/rank
Djibouti	14	27
Equatorial Guinea		
Eritrea		
Eswatini	35.5	30
Ethiopia	16	10
Gabon	12.5	7
The Gambia	15	22.5
Ghana	2	6
Guinea	24	12
Guinea-Bissau		
Kenya	10	19.5
Lesotho	39	26
Liberia	30	22.5
Madagascar	37	40
Malawi	6	21
Mali	28	35
Mauritania	33	14
Mauritius	3	1
Mozambique	34	38
Namibia	20	2
Niger	27	18
Nigeria	41	41
Rwanda	9	8.5
Sao Tome and Principe	1	5
Senegal	12.5	29
Seychelles		
Sierra Leone	22	15
Somalia		
South Africa	7	11
South Sudan		
Sudan	29	37
Tanzania	11	16.5
Togo	25	33
Uganda	17	16.5
Zambia	35.5	19.5
Zimbabwe	8	3

Chapter 5
Modern Slavery, Environmental Destruction and Climate Change

Modern slavery and climate change are current crises. It is stated in [32] that while these phenomena have been well-established separately in recent academic and policy debates, little discussion has taken place on the interconnections between them. The purpose of [32] is to examine these interconnections. It is known that climate change can lead to extreme events which cause an increase in immigration and therefore an increase in modern slavery and human trafficking. However, modern slavery activities may further perpetuating climate change and contributing to significant Greenhouse Gas (GHG) emissions, [32]. The study in [32] focuses on fisheries, fields, forests, and factories.

Modern slavery practices contribute to further environmental destruction and perpetuate climate change. The use of debt-bonded labor in brick kilns, forest clearing and mineral mining damages local environments and ecosystems, while also contributing to climate change through increased levels of GHG emissions. The increasing environmental pressures associated with climate changes in the Global South, e.g., rising sea levels, increasing prevalence and severity of droughts, exacerbate existing vulnerabilities and inequalities of local populations that may render them susceptible to exploitive labor practices, [32].

It is stated in [32] that much of the slave-based and environmentally-destructive activity is enacted to serve the interests of consumers in the Global North who seek cheap and abundant produce, tied up with associated economic pressures in the Global South.

J. N. Mordeson and S. Mathew, *Mathematics of Uncertainty for Coping with World Challenges*, Studies in Systems, Decision and Control 353,
https://doi.org/10.1007/978-3-030-68684-0_5

5.1 Guiasu and Dempster

It is stated in [32] that it is important to understand the nexus of climate change and modern slavery with respect to the Sustainable Development Goals. Four of the most important such SDGs are G_{12}, G_{13}, G_{14}, and G_{15}. This can be seen by the explanation of these SDGs given in the Appendix and summarized from the following table taken from [32, p. 11].

SDG	Primary aim	Relevant sectors
G_{12}	Ensure sustainable consumption and production patterns	Fisheries, Fields, Forests, Factory
G_{13}	Take urgent action to combat climate change and its impacts	Fisheries, Fields, Forests
G_{14}	Conserve and sustainable use the oceans, seas and marine resources for sustainable development	Fisheries
G_{15}	Protect, restore and promote sustainable use of terrestrial ecosystems, sustainably manage forests, combat desertification, and halt and reverse land degradation and halt biodiversity loss	Forests, Fields

We next determine equations that can be used to measure how well countries are doing in meeting the SDGs G_{12}, G_{13}, G_{14}, and G_{15}. The equations are determined from the following table which is a subtable of Table 2.2.1 in [113].

AHP	Bounded difference	Algebraic product	Standard intersection	Row average
G_{12}	0.78	0.79	0.82	0.7967
G_{13}	0.88	0.88	0.88	0.88
G14	0.48	0.52	0.64	0.5467
G_{15}	0.15	0.30	0.45	0.3167
Col. Total	2.29	2.49	2.80	2.5401

Dividing the Row Average entries by 2.5401, we obtain the following AHP equation

$$G = 0.32G_{12} + 0.35G_{13} + 0.22G_{14} + 0.12G_{15}$$

Guiasu/Dempster	Bounded difference	Algebraic product	Standard intersection	Row average	Row product
G_{12}	0.34	0.32	0.29	0.32	0.31552
G_{13}	0.38	0.35	0.31	0.35	0.04123
G_{14}	0.21	0.21	0.23	0.22	0.010143
G_{15}	0.07	0.12	0.16	0.12	0.001344
Col. Total					0.084269

The Row Average entries give the Guiasu equation:

$$G' = 0.32G_{12} + 0.35G_{13} + 0.22G_{14} + 0.12G_{15}.$$

The Row Product entries divided by their Column Total yields the Dempster equation

$$G^d = 0.37G_{12} + 0.49G_{13} + 0.12G_{14} + 0.02G_{15}.$$

Yen	Bounded difference	Algebraic product	Standard intersection	Row average
G_{12}	0.89	0.91	0.94	0.91
G_{13}	1.00	1.00	1.00	1.00
G14	0.55	0.60	0.74	0.63
G_{15}	0.19	0.34	0.52	0.35
Col. Total				2.89

Dividing the Row Average entries by 2.89, we obtain the following Yen equation:

$$G'' = 0.32G_{12} + 0.34G_{13} + 0.21G_{14} + 0.13G_{15}$$

Since the AHP and Guiasu equations are the same and the Yen equation is almost the same, we use only the Guiasu equation to determine the country rankings. We also use the Dempster equation. The entries from the achievement tables in Appendix 1 are substituted into the variables in the above equations to find the achievement score for the countries.

Rankings

Hungary, Mexico, Chile and the Czech Republic rank the highest in achieving the SDGs pertinent to modern slavery related to fisheries, fields, forests, and factories (Tables 5.1, 5.2, 5.3).

Nepal, Philippines, and Pakistan rank the highest in achieving the SDGs pertinent to modern slavery related to fisheries, fields, forests, and factories. Economic pressures associated with depleted fish stocks have fueled the use of slave labor in the Thai fishing industry, [32].

Moldova ranks the highest in achieving the SDGs pertinent to modern slavery related to fisheries, fields, forests, and factories (Tables 5.4, 5.5).

Nicaragua ranks the highest in achieving the SDGs pertinent to modern slavery related to fisheries, fields, forests, and factories. The use of debt-bonded labor in deforestation and forest degradation practices has been primarily investigated in the contexts of Brazil and Indonesia, through focusing on industries such as charcoal, cattle, and palm oil, [32].

Yemen Republic and Egypt rank the highest in achieving the SDGs pertinent to modern slavery related to fisheries, fields, forests, and factories (Table 5.6).

Guinea and Liberia rank the highest in achieving the SDGs pertinent to modern slavery related to fisheries, fields, forests, and factories.

Table 5.1 OECD ranking

OECD		
Country	Guiasu score/rank	Dempster score/rank
Australia	0.426/36	0.395/36
Austria	0.668/19	0.677/22
Belgium	0.603/29	0.633/30
Canada	0.598/30	0.605/32
Chile	0.773/3	0.824/3
Czech Rep.	0.815/2	0.813/4
Denmark	0.680/15	0.702/18
Estonia	0.765/5	0.749/9
Finland	0.619/28	0.611/31
France	0.700/11	0.713/17
Germany	0.649/21	0.682/21
Greece	0.633/27	0.636/28
Hungary	0.832/1	0.841/1
Iceland	0.587/31	0.671/24
Ireland	0.679/17	0.701/19
Isreal	0.549/34	0.635/29
Italy	0.645/23	0.672/23
Japan	0.689/14	0.727/15
Korea Rep.	0.692/12	0.742/10
Latvia	0.740/8	0.761/6
Lithuania	0.749/6	0.755/8
Luxembourg	0.575/32	0.571/33
Mexico	0.772/4	0.828/2
Netherlands	0.634/26	0662/25
New Zealand	0.660/20	0.717/16
Norway	0.504/35	0.472/35
Poland	0.747/7	0.781/5
Portugal	0.691/13	0.728/13
Slovak Rep.	0.734/9	0.723/14
Slovenia	0.679/16	0.728/12
Spain	0.700/10	0.739/11
Sweden	0.649/22	0.686/20
Switzerland	0.639/25	0.646/27
Turkey	0.668/18	0.757/7
U.K.	0.642/24	0.657/26
U.S.	0.569/33	0.547/34

Table 5.2 South and East Asia ranking

South and East Asia		
Country	Guiasu score/rank	Dempster score/rank
Bangladesh	0.809/4	0.885/5
Bhutan	0.780/8	0.867/9
Brunei	0.625/18	0.706/17
Cambodia	0.757/13	0.864/10
China	0.731/14	0.810/14
India	0.799/5	0.884/6
Indonesia	0.776/9	0.870/8
Korea, Dem. Rep.		
Lao PDR	0.793/6	0.890/4
Malaysia	0.707/15	0.783/15
Maldives	0.675/16	0.740/16
Mongolia	0.645/17	0.672/18
Myanmar	0.789/7	0.883/7
Nepal	0.836/1	0.916/1
Pakistan	0.817/3	0.895/3
Philippines	0.827/2	0.897/2
Singapore	0.395/19	0.425/19
Sri Lanka	0.763/11	0.820/13
Thailand	0.775/10	0.831/12
Timor-Leste		
Vietnam	0.762/12	0.849/11

Table 5.3 East Europe and Central Asia ranking

South and East Asia		
Country	Guiasu score/rank	Dempster score/rank
Afghanistan	0.800/4	0.907/2
Albania	0.767/11	0.824/10
Andorra		
Armenia	0.784/7	0.868/5
Azerbaijan	0.725/16	0.813/13
Belarus	0.769/9	0.826/8
Bosnia and Herzegovina	0.625/20	0.706/18
Bulgaria	0.768/10	0.773/16
Croatia	0.814/2	0.836/7
Cyprus	0.546/23	0.563/23

(continued)

Table 5.3 (continued)

South and East Asia		
Country	Guiasu score/rank	Dempster score/rank
Georgia	0.742/14	0.815/12
Kazakhstan	0.628/19	0.669/20
Kyrgz Rep.	0.796/5	0.870/4
Liechtenstein		
Malta	0.662/18	0.700/19
Moldova	0.825/1	0.911/1
Monaco		
Montenegro	0.567/22	0.654/21
N. Macedonia	0.733/15	0.786/15
Romania	0.774/8	0.818/11
Russian Fed.	0.675/17	0.723/17
San Marino		
Serbia	0.750/12	0.786/14
Tajikistan	0.805/3	0.887/3
Turkmenistan	0.573/21	0.618/22
Ukraine	0.746/13	0.825/9
Uzbekistan*	0.787/6	0.866/6

Table 5.4 South America and the Caribbean ranking

South America and the Caribbean		
Country	Guiasu score/rank	Dempster score/rank
Antigua and Baruda		
Argentina	0.720/17	0.802/16
Bahamas, The		
Barbados		
Belize	0.665/21	0.757/19
Bolivia	0.789/8	0.887/2
Brazil	0.777/9	0.829/10
Columbia	0.813/7	0.859/5
Costa Rica	0.816/6	0.856/7
Cuba	0.696/19	0.699/21
Dominica		
Dominican Rep.	0.818/5	0.850/8
Ecuador	0.818/4	0.870/4
El Salvador	0.727/14	0.810/14
Grenada		
Guatimala	0.752/13	0.842/9

(continued)

Table 5.4 (continued)

South America and the Caribbean		
Country	Guiasu score/rank	Dempster score/rank
Guyana	0.618/22	0.639/22
Haiti	0.724/15	0.826/11
Honduras	0.829/3	0.882/3
Jamaica	0.703/18	0.800/17
Nicaragua	0.853/1	0.897/1
Panama	0.755/12	0.817/12
Paraguay*	0.724/16	0.820/12
Peru	0.829/2	0.857/6
Sts. Kitts and Nevis		
St Lucia		
St Vincent and the Grenadines		
Suriname	0.0.767/11	0.783/18
Trinidad and Tobago	0.0.599/23	0.595/23
Urguay	0.683/20	0.755/20
Venezuelia, RB	0.768/10	0.804/15

Table 5.5 Middle East and North Africa Ranking

Middle East and North Africa		
Country	Guiasu score/rank	Dempster score/rank
Algeria	0.767/7	0.845/4
Bahrain	0.578/12	0.571/13
Egypt	0.800/3	0.871/2
Iran	0.804/2	0.836/5
Iraq	0.708/9	0.818/9
Jordan	0.768/6	0.833/7
Kuwait	0.385/16	0.387/16
Lebanon	0.655/11	0.714/10
Libya		
Morocco	0.776/5	0.831/8
Oman	0.671/10	0.695/11
Qatar	0.527/14	0.526/14
Saudi Arabia	0.564/13	0.579/12
Syria	0.728/8	0.835/6
Tunesia	0.796/4	0.850/3
UAE	0.435/15	0.399/15
Yemen Rep.	0.843/1	0.922/1

Table 5.6 Sub-Saharan Africa ranking

Sub-Saharan Africa		
Country	Guiasu score/rank	Dempster score/rank
Angola	0.793/30	0.862/31
Benin	0.847/11	0.904/16
Botswana	0.624/41	0.656/41
Burkino Faso	0.809/27	0.863/30
Burindi	0.842/14	0.917/9
Cabo Verde	0.766/31	0.846/32
Cameroon	0.831/19	0.905/15
C. African Rep.	0.843/13	0.919/8
Chad	0.758/33	0.795/35
Comoros	0.714/34	0.823/33
Congo Dem. Rep.	0.755/35	0.868/29
Congo Rep.	0.862/5	0.900/18
Cote d'Ivoire	0.858/6	0.924/4
Djbouti	0.693/39	0.823/34
Equatorial Guinea		
Eritrea		
Estwatini	0.617/42	0.644/42
Ethiopia	0.810/26	0.900/19
Gabon	0.855/7	0.894/22
The Gambia	0.841/15	0.892/24
Ghana	0.843/12	0.909/10
Guinea	0.891/1	0.938/1
Guinea-Bissau		
Kenya	0.820/25	0.891/26
Lesotho	0.679/40	0.724/39
Liberia	0.887/2	0.955/2
Madagascar	0.822/24	0.904/17
Malawi	0.707/37	0.783/37
Mali	0.830/20	0.909/13
Mauritania	0.765/32	0.794/36
Mauritius	0.576/43	0.630/43
Mozambique	0.873/3	0.919/7
Namibia	0.697/38	0.683/40
Niger	0.806/29	0.875/28
Nigeria	0.836/17	0.899/20
Rwanda	0.826/22	0.909/12
Sao Tome and Principe	0.852/10	0.897/21
Senegal	0.832/18	0.894/23
Seychelles		
Sierra Leone	0.854/9	0.922/5
Somolia		
South Africa	0.713/36	0.760/38
South Sudan		
Sudan	0.854/8	0.921/6
Tanzania	0.870/4	0.933/3
Togo	0.839/16	0.906/14
Uganda	0.827/21	0.909/11
Zambia	0.809/28	0.885/27
Zimbabwe	0.825/23	0.892/25

Similarity Rankings

We next determine the similarity rankings for each region. In the following, when a ranking has missing data for a country, we delete the country for a different ranking even if it has data for the ranking. After the deletion, we rerank. We let μ denote the ranking for the Guiasu method and ν the ranking for the Dempster method.

OECD

We have $M(\mu, \nu) = \frac{624}{708} = 0.881$ and $n = 36$. The smallest possible similarity measure for M with 36 countries and no ties in the ranking can be obtained by substituting 36 for n in the formula $\frac{n+2}{3n+2}$. One obtains $\frac{36+2}{108+2} = 0.0.345$.

We have $S(\mu, \nu) = 1 - \frac{84}{1232} = 0.932$ and $n = 36$. The smallest possible similarity measure for S with 36 countries and no ties in the ranking can be obtained by substituting 36 for n in the formula $\frac{n/2+1}{n+1}$. One obtains $\frac{18+1}{36+1} = 19/37 = 0.514$.

Note that $\frac{M(\mu,\nu)-0.345}{1-0.345} = \frac{0.881-0.345}{1-0.345} = \frac{0.536}{0.655} = 0.818$ and $\frac{S(\mu,\nu))-0.514}{1-0.514} = \frac{0.932-0.514}{1-0.514} = \frac{0.418}{0.486} = 0.860$.

South and East Asia

We have $M(\mu, \nu) = \frac{182}{198} = 0.919$ and $n = 19$. We have $\frac{n+1}{3n-1} = \frac{19+1}{57-1} = \frac{20}{56} = 0.357$.

We have $S(\mu, \nu) = 1 - \frac{16}{380} = 0.958$ and $n = 19$. We have $\frac{1}{2} + \frac{1}{2n} = \frac{1}{2} + \frac{1}{38} = 0.526$.

We have that $\frac{M(\mu,\nu)-0.357}{1-0.357} = \frac{0.919-0.357}{1-0.357} = \frac{0.562}{0.643} = 0.874$ and $\frac{S(\mu,\nu))-0.526}{1-0.526} = \frac{0.958-0.526}{1-0.526} = \frac{0.432}{0.474} = 0.911$.

East Europe and Central Asia

We have $M(\mu, \nu) = \frac{257}{295} = 0.871$ and $n = 23$. We have $\frac{n+1}{3n-1} = \frac{23+1}{69-1} = \frac{24}{68} = 0.353$.

We have $S(\mu, \nu) = 1 - \frac{38}{552} = 0.931$ and $n = 23$. We have $\frac{1}{2} + \frac{1}{2n} = \frac{1}{2} + \frac{1}{46} = 0.522$.

We have that $\frac{M(\mu,\nu)-0.353}{1-0.353} = \frac{0.871-0.353}{1-0.353} = \frac{0.518}{0.647} = 0.801$ and $\frac{S(\mu,\nu))-0.522}{1-0.522} = \frac{0.931-0.522}{1-0.522} = \frac{0.409}{0.478} = 0.856$.

South America and the Caribbean

We have $M(\mu, \nu) = \frac{252}{300} = 0.840$ and $n = 23$. We have $\frac{n+1}{3n-1} = \frac{23+1}{69-1} = \frac{24}{68} = 0.353$.

We have $S(\mu, \nu) = 1 - \frac{48}{552} = 0.913$ and $n = 23$. We have $\frac{1}{2} + \frac{1}{2n} = \frac{1}{2} + \frac{1}{46} = 0.522$.

We have that $\frac{M(\mu,\nu)-0.353}{1-0.353} = \frac{0.840-0.353}{1-0.353} = \frac{0.487}{0.647} = 0.753$ and $\frac{S(\mu,\nu))-0.522}{1-0.522} = \frac{0.913-0.522}{1-0.522} = \frac{0.391}{0.478} = 0.818$.

Middle East and North Africa

We have $M(\mu, \nu) = \frac{127}{145} = 0.876$ and $n = 16$. We have $\frac{n+2}{3n+2} = \frac{16+2}{48+2} = \frac{18}{50} = 0.360$.

We have $S(\mu, \nu) = 1 - \frac{18}{272} = 0.933$ and $n = 16$. We have $\frac{n/2+1}{n+1} = \frac{8+1}{17} = \frac{9}{17} = 0.529$.

We have that $\frac{M(\mu,\nu)-0.360}{1-0.360} = \frac{0.876-0.360}{1-0.360} = \frac{0.516}{0.640} = 0.806$ and $\frac{S(\mu,\nu))-0.529}{1-0.529} = \frac{0.933-0.529}{1-0.529} = \frac{0.404}{0.471} = 0.858$.

Sub-Saharan Africa

We have $M(\mu, \nu) = \frac{863}{1029} = 0837$ and $n = 43$. We have $\frac{n+1}{3n+2} = \frac{43+1}{129+2} = \frac{44}{131} = 0.336$.

We have $S(\mu, \nu) = 1 - \frac{166}{1892} = 0.912$ and $n = 43$. We have $\frac{1}{2} + \frac{1}{2n} = \frac{1}{2} + \frac{1}{86} = 0.512$.

We have that $\frac{M(\mu,\nu)-0.336}{1-0.336} = \frac{0.837-0.336}{1-0.336} = \frac{0.501}{0.664} = 0.755$ and $\frac{S(\mu,\nu))-0.512}{1-0.512} = \frac{0.912-0.512}{1-0.512} = \frac{0.400}{0.488} = 0.820$.

5.2 Stakeholder

In Sect. 2.1, we have that the coefficients for G_{12}, G_{13}, G_{14}, and G_{15} in the Stakeholder equation are 0.12, 0.13, 0.08, and 0.05, respectively. Their sum is 0.38. We divide these coefficients by 0.38 to obtain the following Stakeholder equation

$$SDG = 0.32G_{12} + 0.34G_{13} + 0.21G_{14} + 0.13G_{15}.$$

In [149], countries were assigned colors as a ranking in their achievement of the SDGs. The colors assigned were green, yellow, orange, and red. A green rating on the SDGs denoted achievement, and is assigned to a country for a given SDG only if all the indicators under the goal are rated green; yellow, orange, and red indicate increasing distance from SDG achievement. The rankings of SDGs were determined by averaging the two worst ratings. In order to place the analysis in a fuzzy logic setting, the numbers 0.8, 0.6, 0.4, and 0.2 were assigned to the colors green, yellow, orange, and red, respectively. We also used the aggregation operator average to obtain the average of the color ratings. This gives a method of determining how well a country is doing over all in achieving the SDGs. The achievement tables can be found in Appendix 2.

Rankings

Hungary, Mexico, and the Czech Republic rank the highest in achieving the SDGs pertinent to modern slavery related to fisheries, fields, forests, and factories (Tables 5.7, 5.8).

Bangladesh ranks the highest in achieving the SDGs pertinent to modern slavery related to fisheries, fields, forests, and factories. In Thailand, fishing vessels were found to be making use of debt-bonded labor practices in their operations as a way of reducing labor costs and remaining competitive in the industry, [32]. Migrants from neighboring countries, primarily Cambodia and Myanmar, are exploited and coerced into labor usage (Table 5.9).

Moldova ranks the highest in achieving the SDGs pertinent to modern slavery related to fisheries, fields, forests, and factories (Table 5.10).

Columbia and Nicaragua rank the highest in achieving the SDGs pertinent to modern slavery related to fisheries, fields, forests, and factories (Table 5.11).

Egypt ranks the highest in achieving the SDGs pertinent to modern slavery related to fisheries, fields, forests, and factories (Table 5.12).

Sierra Leone and Tanzania rank the highest in achieving the SDGs pertinent to modern slavery related to fisheries, fields, forests, and factories.

Table 5.7 OECD ranking

OECD		
Country	Color score/rank	Average score/rank
Australia	0.268/27	0.471/22
Austria	0.315/14	0.412/33
Belgium	0.252/32	0.479/18
Canada	0.268/27	0.443/29
Chile	0.332/9	0.531/5
Czech Rep.	0.379/2	0.538/4
Denmark	0.320/12	0.525/9
Estonia	0.370/4	0.607/2
Finland	0.294/19.5	0.489/14
France	0.294/19.5	0.478/20
Germany	0.252/32	0.458/24
Greece	0.294/19.5	0.418/32
Hungary	0.405/1	0.612/1
Iceland	0.226/35	0.408/34
Ireland	0.294/19.5	0.528/6
Isreal	0.200/36	0.391/35
Italy	0.252/32	0.450/26
Japan	0.294/19.5	0.514/12
Korea Rep.	0.332/9	0.522/10
Latvia	0.320/12	0.515/11
Lithuania	0.358/5.5	0.527/7
Luxembourg	0.289/23.5	0.442/30
Mexico	0.374/3	0.547/3
Netherlands	0.320/12	0.488/15
N. Zealand	0.332/9	0.508/13
Norway	0.310/15	0.437/31
Poland	0.278/25	0.525/8
Portugal	0.252/32	0.485/16
Slovak Rep.	0.289/23.5	0.475/21
Slovenia	0.252/32	0.463/23
Spain	0.358/5.5	0.482/17
Sweden	0.294/19.5	0.479/19
Switzerland	0.356/7	0.451/25
Turkey	0.264/29	0.446/27
U.K.	0.268/27	0.445/28
U.S.	0.302/16	0.363/36

Table 5.8 South and East Asia ranking

South and East Asia		
Country	Color score/rank	Average score/rank
Bangladesh	0.596/2	0.655/2
Bhutan	0.508/8	0.618/8
Brunei Darussalam	0.332/15	0.552/16
Cambodia	0.490/10	0.595/13
China	0.290/18.5	0.534/18
India	0.532/6	0.640/5
Indonesia	0.506/9	0.640/4
Korea, Dem. Rep.	0,284/20	0.605/11
Lao PDR	0.580/3.5	0.610/10
Malaysia	0.306/17	0.559/15
Maldives	0.358/14	0.490/19
Mongolia	0.310/16	0.484/20
Myanmar	0.528/7	0.600/12
Nepal	0.580/3.5	0.611/9
Pakistan	0.558/5	0.663/1
Philippines	0.464/12	0.636/7
Singapore	0.226/21	0.440/21
Sri Lanka	0.468/11	0,645/3
Thailand	0.290/18.5	0.539/17
Timor-Leste	0.622/1	0.637/6
Vietnam	0.396/13	0.579/14

Table 5.9 Eastern Europe and Central Asia ranking

Eastern Europe and Central Asia		
Country	Color score/rank	Average score/rank
Afghanistan	0.550/2	0.565/12
Albania	0.448/12	0.611/6
Andorra		
Armenia	0.508/5	0.632/4
Azerbaijan	0.460/10	0.581/8
Belarus	0.460/10	0.540/15.5
Bosnia and Herzegovina	0.354/18	0.524/17
Bulgaria	0.388/14	0.572/11
Croatia	0.468/8	0.621/5
Cyprus	0.252/22	0.383/24
Georgia	0.490/7	0.541/14
Kazakhstan	0.208/23	0.540/15.5

(continued)

Table 5.9 (continued)

Eastern Europe and Central Asia		
Country	Color score/rank	Average score/rank
Kyrgyz Rep.	0.508/5	0.642/3
Liehenstein		
Malta	0.362/16	0.457/21
Moldova	0.584/1	0.674/1
Monaco		
Montenegro	0.264/21	0.412/23
North Macedonia	0.460/10	0.520/18
Romania	0.414/14	0.576/10
Russian Federation	0.332/20	0.514/19
San Marino		
Serbia	0.354/18	0.546/13
Tajikstan	0.542/3	0.671/2
Turmenistan	0.354/18	0.473/20
Ukraine	0.426/12	0.580/9
Uzbekistan	0.508/5	0.697/7

Table 5.10 South America and the Caribbean ranking

South America and the Caribbean		
Country	Color score/rank	Average score/rank
Antigua and Barbuda	0.268/29.5	0.410/ 29
Argentina	0.358/24.5	0.556/13
Bahamas, The	0.268/19.5	0.324/31
Barbados	0.226/31	0.386/30
Belize	0.374/23	0.500/22
Bolivia	0.430/15	0.526/19
Brazil	0.400/21	0.582/9
Columbia	0.574/2	0.672/1
Costa Rica	0.536/4	0.641/6
Cuba	0.422/17.5	0.559/12
Dominica	0.422/17.5	0.507/21
Dominican Rep.	0.506/7	0.643/4
Ecuador	0.574/2	0.642/5
El Salvador	0.422/17.5	0.554/15
Grenada	0.510/6	0.532/18

(continued)

Table 5.10 (continued)

South America and the Caribbean		
Country	Color score/rank	Average score/rank
Guatamala	0.422/11.5	0.612/7
Guyana	0.400/21	0.487/25
Haiti	0.464/9.5	0.586/8
Honduras	0.532/5	0.652/3
Jamaica	0.422/17.7	0.516/20
Nicaragua	0.574/2	0.669/2
Panama	0.464/9.5	0.556/14
Paraguay	0.435/14	0.536/17
Peru	0.442/11.5	0.574/11
St. Kitts and Nevas	0.333/27	0.413/28
St. Luca	0.426/13	0.472/27
St Vincent and Grenadines	0.400/21	0.488/24
Suriname	0.468/8	0.542/16
Trinidad and Tobago	0.332/28	0.475/26
Uruguay	0.336/26	0.488/23
Venezuela, RB	0.358/24.5	0.579/10

Table 5.11 Middle East and North Africa ranking

Middle East and North Africa		
Country	Color score/rank	Average score/rank
Algeria	0.422/8	0.623/2
Bahrain	0.374/9	0.502/12
Egypt, Arab Rep.	0.558/1	0.672/1
Iran, Islamic Rep.	0.468/4	0.613/6
Iraq	0.328/12	0.498/15
Jordan		0.623/3
Kuwait	0.268/15	0.434/16
Lebanon	0.332 / 11	0.532/8
Libya	0.426/7	0.516/11
Morocco	0.448/6	0.611/5
Oman	0.268/15	0.526/9
Qatar	0.336/10	0.482/14
Saudi Arabia	0.268/15	0.485/13
Syrian Arab Rep.	0.464/5	0.597/7
Tunesia	0.532/2.5	0.622/4
United Arab Emirates	0.310/13	0.419/17
Yemen Rep.	0.532/2.5	0.520/10

Table 5.12 Sub-Saharan Africa ranking

Sub-Saharan Africa		
Country	Color score/rank	Average score/rank
Angola	0.380/42	0.594/38
Benin	0.626/7.5	0.658/18
Botswana	0.387/41	0.466/45
Burkino Faso	0.566/21	0.647/23
Burundi	0.686/2.5	0.710/3.5
Cabo Verde	0.554/26.5	0.628/31
Cameroon	0.558/23.5	0.635/28
C. African Rep.	0.686/2.5	0.710/3.5
Chad	0.610/12	0.633/29
Comoros	0.490/34.5	0.539/42
Congo	0.614/10	0.637/27
Congo Rep.	0.545/29	0.654/20
Cote d'Ivoire	0.596/13.5	0.699/6
Djibouti	0.396/40	0.541/41
E. Guinea		0.336/48
Entrea	0.558/23.5	0.674/12
Eswatini	0.353/44	0.519/44
Ethiopia	0.580/16	0.617/34
Gabon	0.558/23.5	0.666/16
The Gambia	0.558/23.5	0.6851/11
Ghana	0.464/37	0.599/37
Guinea	0.638/6	0.707/5
Guinea-Bissau	0.554/26.5	0.609/36
Kenya	0.522/32	0.654/19
Lesotho	0.460/39	0.575/40
Liberia	0.596/13.5	0.698/7
Madagacar	0.570/20	0.644/26
Malawi	0.580/17.5	0.628/32
Mali	0.686/2.5	0.692/9
Mauritania	0.490/34.5	0.653/22
Mauritius	0.242/46.5	0.425/46
Mozambique	0.528/30.5	0.647/25
Nambia	0.358/43	0.613/35
Niger	0.537/30	0.621/33
Nigeria	0.622/9	0.686/10
Rwanda	0.686/10.5	0.661/17
Sao Torre and Pricipe	0.574/19	0.667/15
Senegal	0.516/33	0.646/24
Seychelles	0.242/46.5	0.413/47
Sierra Leone	0.682/2.5	0.756/2
Somalia	0.464/37	0.587/39
South Africa	0.332/45	0.524/43
South Sudan		
Sudan	0.528/30.5	0.654/21
Tanzania	0.664/5	0.737/1
Togo	0.626/7.5	0.693/8
Uganda	0.584/13	0.672/13
Zambia	0.464/37	0.632/30
Zimbabwe	0.580/17.5	0.667/14

Similarity Rankings

We next determine the similarity rankings for each region.

In the following, when a ranking has missing data for a country, we delete the country for a different ranking even if it has data for the ranking. After the deletion, we rerank. We let μ denote the ranking for the Guiasu method and ν the ranking for the Dempster method.

OECD

We have $M(\mu, \nu) = \frac{647.5}{784.5} = 0.698$ and $n = 36$. The smallest possible similarity measure for M with 36 countries and no ties in the ranking can be obtained by substituting 36 for n in the formula $\frac{n+2}{3n+2}$. One obtains $\frac{36+2}{108+2} = 0.345$.

We have $S(\mu, \nu) = 1 - \frac{237}{1332} = 0.822$ and $n = 36$. The smallest possible similarity measure for S with 36 countries and no ties in the ranking can be obtained by substituting 36 for n in the formula $\frac{n/2+1}{n+1}$. One obtains $\frac{18+1}{36+1} = 19/37 = 0.514$.

Note that $\frac{M(\mu,\nu)-0.345}{1-0.345} = \frac{0.698-0.345}{1-0.345} = \frac{0.353}{0.655} = 0.539$ and $\frac{S(\mu.\nu))-0.514}{1-0.514} = \frac{0.822-0.514}{1-0.514} = \frac{0.308}{0.486} = 0.634$.

South and East Asia

We have $M(\mu, \nu) = \frac{195}{267} = 0.730$ and $n = 21$. We have $\frac{n+1}{3n-1} = \frac{22}{62} = 0.355$.

We have $S(\mu, \nu) = 1 - \frac{72}{462} = 0.844$ and $n = 21$. We have $\frac{1}{2} + \frac{1}{2n} = \frac{1}{2} + \frac{1}{42} = 0.524$.

We have that $\frac{M(\mu,\nu)-0.355}{1-0.355} = \frac{0.730-0.355}{1-0.355} = \frac{0.375}{0.645} = 0.581$ and $\frac{S(\mu.\nu))-0.524}{1-0.524} = \frac{0.844-0.524}{1-0.524} = \frac{0.320}{0.476} = 0.672$.

East Europe and Central Asia

We have $M(\mu, \nu) = \frac{234.5}{317.5} = 0.739$ and $n = 23$. We have $\frac{n+1}{3n-1} = \frac{24}{68} = 0.353$.

We have $S(\mu, \nu) = 1 - \frac{87}{552} = 0.842$ and $n = 23$. We have $\frac{1}{2} + \frac{1}{2n} = \frac{1}{2} + \frac{1}{46} = 0.522$.

We have that $\frac{M(\mu,\nu)-0.353}{1-0.353} = \frac{0.739-0.353}{1-0.353} = \frac{0.386}{0.647} = 0.597$ and $\frac{S(\mu.\nu))-0.522}{1-0.522} = \frac{0.842-0.522}{1-0.522} = \frac{0.320}{0.486} = 0.658$.

South America and the Caribbean

We have $M(\mu, \nu) = \frac{430}{562} = 0.765$ and $n = 31$. We have $\frac{n+1}{3n-1} = \frac{32}{92} = 0.348$.

We have $S(\mu, \nu) = 1 - \frac{132}{992} = 0.867$ and $n = 31$. We have $\frac{1}{2} + \frac{1}{2n} = \frac{1}{2} + \frac{1}{62} = 0.516$.

We have that $\frac{M(\mu,\nu)-0.348}{1-0.348} = \frac{0.765-0.348}{1-0.348} = \frac{0.417}{0.652} = 0.640$ and $\frac{S(\mu.\nu))-0.516}{1-0.516} = \frac{0.867-0.516}{1-0.516} = \frac{0.351}{0.484} = 0.725$.

Middle East and North Africa

We have $M(\mu, \nu) = \frac{114}{158} = 0.722$ and $n = 16$. We have $\frac{n+2}{3n+2} = \frac{18}{50} = 0.360$.

We have $S(\mu, \nu) = 1 - \frac{44}{72} = 0.838$ and $n = 16$. We have $\frac{n/2+1}{n+1} = \frac{9}{17} = 0.529$.

We have that $\frac{M(\mu,\nu)-0.360}{1-0.369} = \frac{0.722-0.360}{1-0.360} = \frac{0.362}{0.655} = 0.553$ and $\frac{S(\mu,\nu)-0.529}{1-0.529} = \frac{0.838-0.529}{1-0.529} = \frac{0.309}{0.471} = 0.656$.

Sub-Saharan Africa

We have $M(\mu, \nu) = \frac{989}{1267} = 0.781$ and $n = 47$. We have $\frac{n+1}{3n-1} = \frac{48}{140} = 0.343$.

We have $S(\mu, \nu) = 1 - \frac{277.5}{2265.5} = 0.878$ and $n = 47$. We have $\frac{1}{2} + \frac{1}{2n} = \frac{1}{2} + \frac{1}{94} = 0.511$.

We have that $\frac{M(\mu,\nu)-0.343}{1-0.343} = \frac{0.781-0.343}{1-0.343} = \frac{0.438}{0.657} = 0.667$ and $\frac{S(\mu,\nu)-0.511}{1-0.511} = \frac{0.878-0.511}{1-0.511} = \frac{0.367}{0.489} = 0.751$.

Appendix 1: Achievement

See Tables 5.13, 5.14, 5.15, 5.16, 5.17, 5.18

Table 5.13 SDG achievement

OECD				
	G_{12}	G_{13}	G_{14}	G_{15}
Australia	0.409	0.339	0.563	0.478
Austria	0.455	0.843	0.800	0.714
Belgium	0.467	0.829	0.306	0.850
Canada	0.501	0.685	0.595	0.607
Chile	0.725	0.947	0.662	0.593
Czech Rep.	0.708	0.891	0.800	0.910
Denmark	0.498	0.902	0.489	0.872
Estonia	0.587	0.850	0.813	0.905
Finland	0.487	0.710	0.555	0.821
France	0.534	0.864	0.642	0.767
Germany	0.474	0.902	0.405	0.826
Greece	0.394	0.822	0.594	0.787
Hungry	0.710	0.949	0.800	0.873
Iceland	0.506	0.886	0.359	0.345
Ireland	0.463	0.917	0.534	0.824
Israel	0.425	0.912	0.174	0.506
Italy	0.517	0.847	0.411	0.829
Japan	0.556	0.904	0.536	0.700
Korea Rep.	0.635	0.877	0.548	0.572
Latvia	0.679	0.878	0.509	0.922
Lithuania	0.674	0.841	0.625	0.904
Luxembourg	0.239	0.787	0.700	0.623
Mexico	0.788	0.906	0.695	0.476
Netherlands	0.440	0.883	0.412	0.832
New Zealand	0.515	0.915	0.570	0.471
Norway	0.305	0.544	0.662	0.632
Poland	0.737	0.892	0.437	0.920
Portugal	0.548	0.915	0.518	0.734
Slovak Rep.	0.650	0.772	0.800	0.869

(continued)

Table 5.13 (continued)

OECD				
	G_{12}	G_{13}	G_{14}	G_{15}
Slovenia	0.608	0.912	0.333	0.825
Spain	0.534	0.933	0.594	0.654
Sweden	0.522	0.872	0.423	0.752
Switzerland	0.279	0.889	0.800	0.577
Turkey	0.738	0.899	0.274	0.533
U.K.	0.429	0.846	0.575	0.737
U.S.	0.365	0.661	0.609	0.769

Table 5.14 SDG achievement

East and South Asia				
	G_{12}	G_{13}	G_{14}	G_{15}
Bangladesh	0.905	0.971	0.519	0.609
Bhutan	0.877	0.970		0.587
Brunei	0.890	0.723	0.087	0.619
Cambodia	0.973	0.927	0.344	0.444
China	0.820	0.920	0.362	0.627
India	0.945	0.945	0.512	0.511
Indonesia	0.910	0.948	0.504	0.419
Korea, Dem. Rep.				
Lao PDR	0.945	0.965		0.516
Malaysia	0.771	0.878	0.492	0.430
Maldives	0.764	0.789	0.524	0.381
Mongolia	0.808	0.645		0.723
Myanmar	0.961	0.950	0.428	0.526
Nepal	0.982	0.980		0.688
Pakistan	0.921	0.987	0.476	0.670
Philippines	0.942	0.944	0.622	0.554
Singapore	0.350	0.505	0.352	0.274
Sri Lanka	0.779	0.923	0.557	0.634
Thailand	0.795	0.935	0.547	0.670
Timor-Leste				
Vietnam	0.871	0.945	0.452	0.486

Table 5.15 SDG achievement

East Europe and Central Asia				
	G_{12}	G_{13}	G_{14}	G_{15}
Afghanistan	0.982	0.970		0.521
Albania	0.832	0.921	0.410	0.800
Andorra				
Armenia	0.901	0.950		0.612
Azerbaijan	0.891	0.906	0.225	0.669
Belarus	0.826	0.923		0.781
Bosnia and Herzegovina	0.890	0.723	0.087	0.619
Bulgaria	0.666	0.877	0.653	0.933
Croatia	0.735	0.936	0.748	0.792
Cyprus	0.417	0.722	0.328	0.775
Geogia	0.822	0.913	0.430	0.606
Kazakhstan	0.642	0.751	0.434	0.586
Kyrgz Rep.	0.917	0.942		0.680
Liechtenstein				
Malta	0.484	0.911	0.502	0.706
Moldova	0.970	0.979		0.641
Monaco				
Montenegro	0.606	0.794	0.284	0.322
N. Macedonia	0.812	0.858		0.740
Romania	0.719	0.952	0.533	0.843
Russian Fed.	0.691	0.822	0.425	0.662
San Marino				
Serbia	0.837	0.850		0.495
Tajikistan	0.932	0.962		0.686
Turkmenistan	0.878	0.515	0.258	0.506
Ukraine	0.804	0.956	0.386	0.635
Uzbekistan	0.921	0.934		0.622

Table 5.16 SDG achievement

Latin America and the Caribbean				
	G_{12}	G_{13}	G_{14}	G_{15}
Antigua and Baruda				
Argentina	0.791	0.921	0.399	0.530
Bahamas, The				
Barbdos				
Belize	0.765	0.873	0.312	0.439
Bolivia	0.876	0.944		0.731
Brazil	0.787	0.917	0.632	0.609
Columbia	0.848	0.907	0.749	0.564
Costa Rica	0.827	0.934	0.662	0.672
Cuba	0.906	0.567	0.601	0.683
Dominica				
Dominican Rep.	0.860	0.889	0.678	0.752
Ecuador	0.848	0.939	0.702	0.601
El Salvador	0.864	0.906	0.278	0.659
Grenada				
Guatemala	0.856	0.949	0.411	0.530
Guyana	0.562	0.720	0.550	0.595
Haiti	0.918	0.894	0.330	0.436
Honduras	0.883	0.944	0.662	0.652
Jamaica	0.908	0.870	0.218	0.558
Nicaragua	0.902	0.949	0.692	0.738
Panama	0.802	0.902	0.554	0.572
Paraguay	0.818	0.930		0.447
Peru	0.788	0.931	0.788	0.713
St. Kitts and Nevis				
St. Lucia				
St. Vincent and the Grenadines				
Suriname	0.798	0.791	0.712	0.713
Trinidad and Tobago	0.752	0.491	0.525	0.641
Uruguay	0.724	0.869	0.422	0.514
Venezuelia, RB	0.785	0.890	0.514	0.789

Table 5.17 SDG achievement

Middle East and North Africa

	G_{12}	G_{13}	G_{14}	G_{15}
Algeria	0.865	0.943	0.419	0.632
Bahrain	0.721	0.457	0.578	0.548
Eygpt	0.829	0.978	0.566	0.634
Iran	0.806	0.890	0732	0.680
Iraq	0.906	0.900	0.285	0.394
Jordan	0.858	0.948	0.279	0.902
Kuwait	0.222	0.502	0.424	0.410
Lebanon	0.756	0.774	0.366	0.566
Libya				
Morocco	0.825	0.924	0.482	0.756
Oman	0.698	0.711	0.651	0.517
Qatar	0.580	0.499	0.455	0.596
Saudi Arabia	0.567	0.597	0.557	0.473
Syria	0.854	0.967	0.300	0.476
Tunesia	0.867	0.907	0.594	0.656
UAE	0.409	0.334	0.625	0.454
Yemen Rep.	0.955	0.984	0.637	0.510

Table 5.18 SDG achievement

Sub-Saharan Africa

	G_{12}	G_{13}	G_{14}	G_{15}
Angola	0.937	0.909	0.478	0.650
Benin	0.949	0.973	0.497	0.847
Botswana	0.674	0.715		0.717
Burkino Faso	0.937	0.908		0.821
Burundi	0.966	0.993		0.739
Cabo Verde	0.918	0.899	0.455	0.547
Cameroon	0.949	0.979	0.502	0.686
C. African Rep.	0.961	0.994		0.899
Chad	0.959	0.763		0.807
Comoros	0.839	0.974	0.283	0.412
Congo, Dem. Rep.	0.947	0.992	0.153	0.654
Congo Rep.	0.931	0.952	0.593	0.911
Cote d'lvoire	0.962	0.994	0.546	0.760
Djbouti	0.952	0.909	0.149	0.370

(continued)

Table 5.18 (continued)

	G_{12}	G_{13}	G_{14}	G_{15}
Equatorial Guinea				
Eritea				
Estwatini	0.897	0.527		0.633
Ethiopia	0.980	0.957		0.567
Gabon	0.911	0.947	0.629	0.855
The Gambia	0.936	0.935	0.604	0.749
Ghana	0.973	0.965	0.510	0.754
Guinea	0.949	0.992	0.708	0.781
Guinea-Bissau				
Kenya	0.931	0.948	0.581	0.592
Lesotho	0.921	0.662		0.6212
Liberia	0.993	0.993	0.752	0.540
Madagascar	0.946	0.969	0.578	0.506
Malawi	0.771	0.878	0.492	0.430
Mali	0.949	0.992		0.699
Mauritania	0.937	0.732	0.625	0.658
Mauritius	0.614	0.680	0.541	0.235
Mozambique	0.987	0.922	0.740	0.666
Namibia	0.794	0.611	0.605	0.855
Niger	0.912	0.953		0.737
Nigeria	0.948	0.961	0.514	0.763
Rwanda	0.958	0.986		0.665
Sao Tome and Principe	0.949	0.896	0.794	0.574
Senegal	0.950	0.948	0.522	0.749
Seychelles				
Sierra Leone	0.957	0.984	0.605	0.656
Somalia				
South Africa	0.688	0.870	0.565	0.591
South Sudan				
Sudan	0.934	0.987	0.664	0.605
Tanzania	0.979	0.982	0.650	0.569
Togo	0.959	0.986	0.419	0.863
Uganda	0.958	0.986		0.672
Zambia	0.897	0.984		0.703
Zimbabwe	0.937	0.964		0.777

Appendix 2: Achievement (Color and Average)

See Tables 5.19, 5.20, 5.21, 5.22, 5.23, 5.24

Table 5.19 Color/Average achievement

OECD								
Color/Average	G_{12}	G_{13}	G_{14}	G_{15}	G_{12}	G_{13}	G_{14}	G_{15}
Australia	0.2	0.2	0.4	0.4	.560	.320	.600	.440
Austria	0.2	0.2	0.5	0.6	.333	.400		.640
Belgium	0.2	0.2	0.2	0.6	.433	.480	.400	.720
Canada	0.2	0.2	0.4	0.4	.440	.320	.550	.600
Chile	0.4	0.2	0.4	0.4	.560	.480	.550	.560
Czech Rep.	0.4	0.2	0.5	0.6	.467	.520		.760
Denmark	0.2	0.4	0.2	0.6	.467	.550	.450	.720
Estonia	0.2	0.2	0.4	0.5	.500	.560	.750	.760
Finland	0.2	0.2	0.4	0.6	.400	.400	.600	.760
France	0.2	0.2	0.4	0.6	.400	.400	.600	.680
Germany	0.2	0.2	0.2	0.6	.400	.480	.350	.720
Greece	0.2	0.2	0.4	0.6	.200	.440	.550	.680
Hungary	0.4	0.2	0.5	0.8	.633	.520		.800
Iceland	0.2	0.2	0.2	0.4	.300	.500	.400	.450
Ireland	0.2	0.2	0.4	0.6	.400	.560	.550	.720
Isreal	0.2	0.2	0.2	0.2	.300	.520	.267	.480
Italy	0.2	0.2	0.2	0.6	.367	.440	.400	.760
Japan	0.2	0.2	0.4	0.6	.467	.520	.500	.640
Korea Rep.	0.4	0.2	0.4	0.4	.480	.560	.500	.560
Latvia	0.2	0.4	0.2	0.6	.433	.600	.350	.760
Lithuania	0.4	0.2	0.4	0.6	.467	.450	.600	.760
Luxembourg	0.2	0.2	0.5	0.4	.300	.500		.640
Mexico	0.4	0.4	0.4	0.2	.567	.520	.600	.480
Netherlands	0.2	0.4	0.2	0.6	.333	.600	.400	.722
N. Zealand	0.4	0.2	0.4	0.4	.480	.520	.550	.480
Norway	0.2	0.2	0.6	0.4	.300	.450	.400	.800
Poland	0.2	0.2	0.2	0.8	.500	.520	.400	.800
Portugal	0.2	0.2	0.2	0.6	.333	.560	.450	.720
Slovak Rep	0.4	0.2	0.5	0.6	.440	.400		.760
Slovenia	0.2	0.2	0.2	0.6	.468	.400	.400	.720
Spain	0.4	0.2	0.4	0.6	.367	.520	.500	.640
Sweden	0.2	0.2	0.4	0.6	.400	.480	.450	.720
Switzerland	0.2	0.4	0.5	0.4	.300	.520		.640
Turkey	0.4	0.2	0.2	0.2	.467	.520	.250	.520
U.K.	0.2	0.2	0.4	0.4	.360	.400	.550	.600
U.S.	0.2	0.2	0.6	0.6	.200	.280	.600	.600

Table 5.20 Color/Average achievement

East and South Asia								
Color/Average	G_{12}	G_{13}	G_{14}	G_{15}	G_{12}	G_{13}	G_{14}	G_{15}
Bangladesh	0.8	0.6	0.4	0.4	0.80	0.65	0.50	0.56
Bhutan	0.6	0.6		0.4	0.70	0.73		0.56
Brunei Darussalam	0.4	0.2	0.4	0.4	0.65	0.47	0.53	0.56
Cambodia	0.6	0.6	0.2	0.4	0.76	0.65	0.30	0.52
China	0.4	0.2	0.2	0.4	0.63	0.50	0.40	0.60
India	0.6	0.6	0.4	0.4	0.77	0.65	0.50	0.52
Indonesia	0.6	0.6	0.4	0.2	0.73	0.75	0.45	0.44
Korea, Dem. Rep.		0.4	0.2	0.4	0.80	0.60	0.40	0.47
Lao PDR	0.8	0.6		0.4	0.80	0.65		0.47
Malaysia	0.4	0.2	0.4	0.2	0.57	0.60	0.50	0.52
Maldives	0.4	0.4	0.2	0.4	0.57	0.47	0.50	0.33
Mongolia	0.4	0.2		0.6	0.67	0.35		0.72
Myanmar	0.8	0.4	0.4	0.4	0.80	0.55	0.45	0.48
Nepal	0.8	0.6		0.4	0.80	0.65		0.68
Pakistan	0.6	0.6	0.4	0.6	0.77	0.75	0.35	0.68
Philippines	0.6	0.4	0.4	0.4	0.73	0.60	0.60	0.56
Singapore	0.2	0.2	0.2	0.4	0.73	0.33	0.20	0.40
Sri Lanka	0.4	0.6	0.4	0.4	0.70	0.70	0.50	0.60
Thailand	0.4	0.2	0.2	0.4	0.63	0.50	0.45	0.56
Timor-Leste	0.8	0.6	0.4	0.6	0.80	0.60	0.47	0.60
Vietnam	0.6	0.4	0.2	0.2	0.68	0.60	0.45	0.48

Table 5.21 Color/Average achievement

Eastern Europe and Central Asia								
Color/Average	G_{12}	G_{13}	G_{14}	G_{15}	G_{12}	G_{13}	G_{14}	G_{15}
Afghanistan	0.8	0.6		0.2	0.80	0.60		0.48
Albania	0.6	0.4	0.2	0.6	0.70	0.60	0.40	0.76
Andorra		0.2		0.4	0.80	0.20		0.44
Armenia	0.6	0.6		0.4	0.70	0.75		0.60
Azerbaijan	0.6	0.4		0.6	0.72	0.55	0.40	0.68
Belarus	0.6	0.4		0.6	0.65	0.50		0.76
Bosnia and Herzegovina	0.6	0.2	0.2	0.4	0.70	0.50	0.20	0.68
Bulgaia	0.2	0.4	0.4	0.8	0.53	0.55	0.53	0.80
Croatia	0.4	0.4	0.6	0.6	0.53	0.65	0.65	0.72
Cyprus	0.2	0.2	0.2	0.6	0.27	0.40	0.33	0.70
Georgia	0.6	0.6	0.2	0.4	0.60	0.55	0.40	0.60

(continued)

Table 5.21 (continued)

Color/Average	G_{12}	G_{13}	G_{14}	G_{15}	G_{12}	G_{13}	G_{14}	G_{15}
Kazakhstan	0.2	0.2		0.4	0.52	0.50	0.60	0.60
Kyrgyz Republic	0.6	0.6		0.4	0.75	0.70		0.68
Liehenstein		0.6		0.4		0.50		0.60
Malta	0.2	0.6	0.2	0.4	0.30	0.60	0.50	0.40
Moldovia	0.6	0.8		0.4	0.75	0.80		0.64
Monaco		0.4				0.50	0.20	0.20
Montenegro	0.4	0.2	0.2	0.2	0.47	0.45	0.27	0.40
North Macedonia	0.6	0.4		0.6	0.65	0.41		0.70
Romania	0.2	0.6	0.2	0.8	0.47	0.70	0.40	0.80
Russian Federation	0.4	0.2	0.4	0.4	0.52	0.50	0.45	0.64
SanMarino		0.2	0.4	0.4		0.20		0.50
Serbia	0.6	0.2		0.4	0.73	0.50		0.60
Tajikistan	0.8	0.5		0.4	0.80	0.73		0.68
Turkmenistan	0.6	0.2		0.4	0.76	0.20	0.40	0.60
Ukraine	0.4	0.6	0.2	0.4	0.60	0.65	0.40	0.64
Uzbekistan	0.6	0.6		0.4	0.75	0.60		0.64

Table 5.22 Color/Average achievement

Latin America and the Caribbean								
Color/Average	G_{12}	G_{13}	G_{14}	G_{15}	G_{12}	G_{13}	G_{14}	G_{15}
Antigua and Barbuda	0.2	0.2	0.4	0.4	0.40	0.35	0.53	0.40
Argentina	0.4	0.4	0.2	0.4	0.63	0.55	0.45	0.56
Bahamas, The	0.2	0.2	0.4	0.4	0.40	0.20	0.33	0.45
Barbados	0.2	0.2	0.2	0.4	0.36	0.45	0.27	0.47
Belize	0.4	0.4	0.4	0.2	0.60	0.55	0.33	0.40
Bolivia	0.6	0.4		0.4	0.73	0.60		0.68
Brazil	0.4	0.4	0.4	0.4	0.63	0.55	0.55	0.60
Columbia	0.6	0.6	0.6	0.4	0.67	0.70	0.70	0.56
Costa Rica	0.4	0.6	0.6	0.6	0.63	0.65	0.70	0.55
Cuba	0.6	0.4	0.2	0.4	0.75	0.40	0.50	0.60
Dominica	0.6	0.4	0.2	0.4	0.60	0.60	0.20	0.53
Dominican Rep.	0.6	0.4	0.6	0.4	0.68	0.65	0.60	0.60
Ecuador	0.6	0.6	0.6	0.4	0.63	0.65	0.70	0.56
El Salvador	0.6	0.4	0.2	0.4	0.67	0.60	0.30	0.56
Grenada	0.4	0.6	0.6	0.4	0.47	0.70	0.50	0.47
Guatamala	0.4	0.6	0.4	0.2	0.70	0.65	0.50	0.48
Guyana	0.4	0.4	0.4	0.4	0.76	0.75	0.60	0.64
Haiti	0.6	0.6	0.2	0.2	0.70	0.65	0.40	0.44
Honduras	0.6	0.6	0.4	0.4	0.73	0.65	0.60	0.55

(continued)

Table 5.22 (continued)

Color/Average	G_{12}	G_{13}	G_{14}	G_{15}	G_{12}	G_{13}	G_{14}	G_{15}
Jamaica	0.6	0.4	0.2	0.4	0.70	0.50	0.27	0.50
Nicaragua	0.6	0.6	0.6	0.4	0.73	0.65	0.60	0.68
Panama	0.6	0.4	0.4	0.4	0.60	0.60	0.45	0.50
Paraguay	0.4	0.6		0.4	0.63	0.60		0.52
Peru	0.4	0.4	0.6	0.4	0.60	0.55	0.60	0.53
St. Kitts and Nevas	0.2	0.6	0.2		0.40	0.60	0.20	0.30
St. Luca	0.4	0.4	0.4	0.6	0.47	0.53	0.30	0.60
St. Vincent and Grenadines	0.4	0.4	0.4	0.4	0.53	0.53	0.33	0.53
Suriname	0.4	0.4	0.6	0.6	0.64	0.30	0.70	0.68
Trinidad and Tobago	0.4	0.2	0.4	0.4	0.56	0.35	0.50	0.55
Uruguay	0.2	0.4	0.4	0.4	0.57	0.50	0.35	0.48
Venezuela, RB	0.4	0.2	0.4	0.6	0.64	0.50	0.55	0.68

Table 5.23 Color/Average achievement

Middle East and North Africa

	G_{12}	G_{13}	G_{14}	G_{15}	G_{12}	G_{13}	G_{14}	G_{15}
Algeria	0.6	0.4	0.2	0.4	0.70	0.67	0.45	0.60
Bahrain	0.4	0.2	0.6	0.4	0.53	0.47	0.53	0.47
Eygpt, Arab Rep.	0.6	0.6	0.4	0.6	0.70	0.75	0.50	0.68
Iran, Islamic Rep.	0.4	0.4	0.6	0.6	0.60	0.53	0.75	0.64
Iraq	0.6	0.2	0.2	0.2	0.70	0.45	0.33	0.40
Jordan	0.6				0.64	0.80	0.20	0.80
Kuwait	0.2	0.2	0.4	0.4	0.30	0.50	0.47	0.53
Lebanon	0.4	0.2	0.4	0.4	0.52	0.60	0.40	0.60
Libya	0.4	0.6	0.2	0.4	0.64	0.50	0.30	0.60
Morocco	0.6	0.4	0.2	0.6	0.70	0.60	0.45	0.68
Oman	0.2	0.2	0.4	0.4	0.47	0.50	0.65	0.53
Qatar	0.2	0.2	0.6	0.6	0.43	0.50	0.50	0.53
Saudi Arabia	0.2	0.2	0.4	0.4	0.43	0.50	0.50	0.56
Syrian Arab Rep.	0.6	0.6	0.2	0.2	0.67	0.73	0.33	0.50
Tunesia	0.6	0.6	0.4	0.4	0.70	0.60	0.55	0.60
United Arab Emirates	0.2	0.2	0.6	0.4	0.32	0.35	0.65	0.47
Yemen, Rep.	0.6	0.6	0.4	0.4	0.33	0.70	0.50	0.55

Table 5.24 Color/Average achievement

Sub-Saharan Africa

Color/Average	G_{12}	G_{13}	G_{14}	G_{15}	G_{12}	G_{13}	G_{14}	G_{15}
Angola	0.6	0.2	0.2	0.6	0.77	0.50	0.45	0.64
Benin	0.6	0.8	0.4	0.6	0.77	0.80	0.20	0.75
Botswana	0.4	0.4		0.6	0.48	0.50		0.68
Burkino Faso	0.6	0.6		0.8	0.77	0.65		0.80
Burundi	0.8	0.8		0.6	0.80	0.80		0.76
Cabo Verde	0.8	0.6	0.2	0.4	0.80	0.65	0.35	0.60
Cameroon	0.6	0.6	0.4	0.6	0.77	0.65	0.40	0.64
C. African Rep.	0.8	0.8		0.6	0.80	0.80		0.76
Chad	0.8	0.6		0.6	0.80	0.60		0.76
Comoros	0.6	0.6	0.2	0.4	0.65	0.60	0.40	0.33
Congo	0.6	0.8		0.6	0.75	0.80	0.20	0.64
Congo Rep.	0.6	0.6	0.4	0.5	0.76	0.67	0.40	0.76
Cote d'Ivoire	0.8	0.6	0.4	0.4	0.80	0.75	0.50	0.64
Djibouti	0.6	0.4	0.2	0.2	0.76	0.60	0.20	0.40
E. Guinea			0.4	0.4		0.20	0.60	0.60
Entrea	0.6	0.8	0.2	0.4	0.77	0.80	0.40	0.55
Eswatini	0.6	0.2		0.4	0.73	0.45		0.55
Ethiopia	0.8	0.6		0.4	0.80	0.65		0.52
Gabon	0.6	0.6	0.4	0.6	0.70	0.70	0.50	0.76
The Gambia	0.6	0.6	0.4	0.6	0.76	0.65	0.60	0.73
Ghana	0.6	0.4	0.4	0.4	0.75	0.60	0.40	0.55
Guinea	0.7	0.7	0.4	0.4	0.76	0.75	0.60	0.64
Guinea-Bissau	0.8	0.6	0.2	0.4	0.80	0.50	0.50	0.60
Kenya	0.6	0.6	0.4	0.4	0.77	0.66	0.55	0.52
Lesotho	0.6	0.4		0.6	0.73	0.55		0.67
Liberia	0.8	0.6	0.4	0.4	0.80	0.70	0.65	0.52
Madagascar	0.8	0.6	0.4	0.2	0.80	0.65	0.50	0.48
Malawi	0.8	0.6		0.4	0.80	0.65		0.60
Mali	0.8	0.8		0.6	0.80	0.80		0.64
Mauritania	0.6	0.4	0.4	0.6	0.77	0.60	0.55	0.67
Mauritius	0.2	0.2	0.4	0.2	0.40	0.40	0.55	0.35
Mozamgique	0.8	0.4	0.4	0.4	0.80	0.55	0.60	0.60
Nambia	0.4	0.2	0.4	0.6	0.60	0.55	0.65	0.75
Niger	0.6	0.6		0.6	0.73	0.65		0.72
Nigeria	0.8	0.6	0.4	0.6	0.80	0.65	0.53	0.75

(continued)

Table 5.24 (continued)

Color/Average	G_{12}	G_{13}	G_{14}	G_{15}	G_{12}	G_{13}	G_{14}	G_{15}
Rwanda	0.6	0.8		0.6	0.73	0.80		0.60
Sao Torre and Principe	0.6	0.6	0.6	0.4	0.73	0.67	0.65	0.53
Senegal	0.6	0.6	0.2	0.6	0.77	0.65	0.45	0.65
Seychelles	0.2	0.2	0.4	0.2	0.50	0.35	0.47	0.27
Sierra Leone	0.7	0.7	0.5	0.5	0.77	0.75	0.60	0.68
Somalia	0.6	0.6	0.2	0.2	0.70	0.65	0.40	0.45
South Africa	0.4	0.2	0.4	0.4	0.57	0.45	0.55	0.56
South Sudan		0.6		0.6		0.60		0.75
Sudan	0.8	0.4	0.4	0.4	0.80	0.60	0.55	0.60
Tanzania	0.8	0.8	0.4	0.4	0.80	0.80	0.65	0.56
Togo	0.6	0.8	0.45	0.6	0.77	0.80	0.40	0.70
Uganda	0.6	0.8		0.4	0.76	0.80		0.60
Zambia	0.4	0.6		0.6	0.67	0.75		0.68
Zimbabwe	0.8	0.6		0.4	0.80	0.75		0.60

Chapter 6
Modern Slavery

Modern slavery is a destructive, personal crime and abuse of human rights. It is a wide spread and profitable criminal industry. Even so, it is largely invisible, due in part because it disproportionately affects the most marginalized. Consequently, measuring the problem is so crucial in exposing and ultimately resolving it. The Global Slavery Index 2018 is an important endeavor in the measurement of the problem. Because of the lack of accurate data, our purpose is to bring to bear the use of mathematics of uncertainty to help analyze the situation.

6.1 Government Response, Vulnerability, and Prevalence

The Global Slavery Index is a tool to aid in the understanding the size of the problem, existing responses, and contributing factors so that they can advocate for a build sound policies that will eradicate modern slavery. In this section we consider the findings in the Global Slavery Index concerning the prevalence of modern slavery in countries, the vulnerability of countries to modern slavery, and the government response to combatting modern slavery.

The following tables are from The Global Slavery Index 2018. They provide regional rankings for countries making up various regions. Our goal is to use fuzzy similarity measures to compare the similarity between the prevalence ranking the vulnerability ranking.

J. N. Mordeson and S. Mathew, *Mathematics of Uncertainty for Coping with World Challenges*, Studies in Systems, Decision and Control 353,
https://doi.org/10.1007/978-3-030-68684-0_6

Africa

Country	Government Response Rank	Vulnerability Rank	Prevalence Rank
South Africa	1	39	47
Senegal	2	46	46
Sierra Leone	3	38	37
Nigeria	4	7	15
Tunisia	5	50	50
Cote d'lvoire	6	34	31
Uganda	7	28	16
Mozambique	8	35	36
Egypt	9	30	34
Lesotho	10	33	41
Benin	11	47	35
Morocco	12	44	49
Kenya	13	10	23
Algeria	14	42	48
Ethiopia	15	19	30
Burkina Faso	16	40	40
Djibouti	17	27	22
Mauritius	18	51	51
Gambia	19	32	32
Rwanda	20	24	10
Nambia	21	45	45
Botswana	22	49	44
Tanzania, U. Rep.	23	29	29
Madagascar	24	23	17
Zambia	25	37	33
Liberia	26	31	20

Africa continued

Country	Government Response Rank	Vulnerability Rank	Prevalence Rank
Mali	27	36	43
Angola	28	21	21
Swaziland	29	18	11
Ghana	30	41	38
Malawi	31	20	18
Niger	32	17	26
Cameroon	33	12	24
Gabon	34	43	39
Togo	35	26	25
Cape Verde	36	48	42
Guinea	37	16	13
Zimbabwe	38	15	27
Congo, Democratic Rep.	39	3	7
Guinea-Bissau	40	11	19
Chad	41	6	9
Somalia	42	4	6
Mauritania	43	22	4
Sudan	44	5	8
Congo	45	14	12
Burundi	46	9	2
Equatorial Guinea	47	25	28
Central African Rep.	48	1	3
Eritrea	49	13	1
Libya	50	8	14
South Sudan		2	5

Similarity Measure Let ν denote the ranking with respect to vulnerability and π the ranking with respect to prevalence.

We find that $M(\nu, \pi) = \frac{1188}{1464} = 0.811$. The smallest possible similarity measure for 51 countries with no ties in the ranking can be obtained by substituting 51 for n in the formula $\frac{n+1}{3n-1}$. One obtains $52/152 = 0.342$. We have that $\frac{M(\nu,\pi)-0.342}{1-0.342} = \frac{0.811-0.342}{1-0.342} = 0.714$.

We also have that $S(\nu, \pi) = 1 - \frac{278}{2662} = 0.896$. The smallest possible similarity measure for 51 countries with no ties can be obtained by substituting 51 for n in the formula $\frac{1}{2} + \frac{1}{2n}$. One obtains $\frac{1}{2} + \frac{1}{102} = 0.51$. We have that $\frac{S(\nu,\pi)-0.51}{1-0.51} = \frac{0.896-0.51}{1-0.51} = 0.788$.

Let μ denote the ranking with respect to government response and ν^r the reverse ranking of the ranking with respect to vulnerability. Then $M(\mu, \nu^r) = \frac{1082}{1570} = 0.689$. The smallest possible similarity measure for 51 countries with no ties in the ranking 0.342. We have that $\frac{M(\mu,\nu^r)-0.342}{1-0.342} = \frac{0.689-0.342}{1-0.342} = 0.527$.

We also have that $S(\mu, \nu^r) = 1 - \frac{488}{2652} = 1 - 0.184 = 0.816$. The smallest possible similarity measure for 51 countries with no ties is 0.51. We have that $\frac{S(\mu,\nu^r)-0.51}{1-0.51} = \frac{0.816-0.51}{1-0.51} = 0.624$.

Americas

Country	Government Response Rank	Vulnerability Rank	Prevalence Rank
United States	1	26	23
Argentina	2	22	22
Chile	3	24	26
Canada	4	27	27
Jamaica	5	11	13
Dominican Rep.	6	14	3
Brazil	7	21	20
Peru	8	10	14
Mexico	9	3	11
Uruguay	10	25	25
Costa Rica	11	23	24
Trinidad and Tobago	12	19	6
Ecuador	13	17	16
Nicaragua	14	13	8
Guatemala	15	6	7
Panama	16	20	19
Bolivia	17	12	18
Columbia	18	7	10
Paraguay	19	18	21
Haiti	20	1	2
Barbados	21	16	9
El Salvador	22	8	15
Honduras	23	4	5
Guyana	24	9	12
Venezuela	25	2	1
Rep.of Suriname	26	15	17
Cuba	27	5	4

Similarity Measure

Let ν denote the ranking with respect to vulnerability and π the ranking with respect to prevalence.

We find that $M(\nu, \pi) = \frac{334}{422} = 0.791$. The smallest possible similarity measure for 27 countries with no ties in the ranking can be obtained by substituting 27 for n in the formula $\frac{n+1}{3n-1}$. One obtains $28/80 = 0.35$. We have that $\frac{M(\nu,\pi)-0.35}{1-0.35} = \frac{0.791-0.35}{1-0.35} = 0.678$.

We also have that $S(\nu, \pi) = 1 - \frac{88}{756} = 0.884$. The smallest possible similarity measure for 27 countries with no ties can be obtained by substituting 27 for n in the formula $\frac{1}{2} + \frac{1}{2n}$. One obtains $\frac{1}{2} + \frac{1}{54} = 0.519$. We have that $\frac{S(\nu,\pi)-0.519}{1-0.519} = \frac{0.884-0.519}{1-0.519} = 0.759$.

Let μ denote the ranking with respect to government response and ν^r the reverse ranking of the ranking with respect to vulnerability. Then $M(\mu, \nu^r) = \frac{306}{450} = 0.680$. The smallest possible similarity measure for 27 countries with no ties in the ranking is $\frac{n+1}{3n-1} = \frac{28}{80}$ 0.35. We have that $\frac{M(\mu,\nu^r)-0.35}{1-0.35} = \frac{0.680-0.35}{1-0.35} = 0.508$.

We also have that $S(\mu, \nu^r) = 1 - \frac{144}{756} = 1 - 0.190 = 0.810$. The smallest possible similarity measure for 27 countries with no ties is $\frac{1}{2} + \frac{1}{2n} = \frac{1}{2} + \frac{1}{54} = 0.519$. We have that $\frac{S(\mu,\nu^r)-0.519}{1-0.519} = \frac{0.810-0.519}{1-0.519} = 0.605$.

Arab States

Country	Government Response Rank	Vulnerability Rank	Prevalence Rank
United Arab Emirates	1	11	9
Jordan	2	6	7
Qatar	3	10	10
Bahrain	4	7	6
Oman	5	5	4
Lebanon	6	4	8
Saudi Arabia	7	8	5
Kuwait	8	9	11
Syrian Arab Republic		1	1
Yemen		2	3
Iraq		3	2

Similarity Measure

Let ν denote the ranking with respect to vulnerability and π the ranking with respect to prevalence. Then

We find that $M(\nu, \pi) = \frac{58}{74} = 0.784$. The smallest possible similarity measure for 11 countries with no ties in the ranking can be obtained by substituting 11 for n in the formula $\frac{n+1}{3n-1}$. One obtains $12/32 = 0.375$. We have that $\frac{M(\nu,\pi)-0.375}{1-0.375} = \frac{0.784-0.375}{1-0.375} = 0.654$.

We also have that $S(\nu, \pi) = 1 - \frac{16}{132} = 0.879$. The smallest possible similarity measure for 11 countries with no ties can be obtained by substituting 11 for n in the formula $\frac{1}{2} + \frac{1}{2n}$. One obtains $\frac{1}{2} + \frac{1}{22} = 0.545$. We have that $\frac{S(\nu,\pi)-0.545}{1-0.545} = \frac{0.879-0.545}{1-0.545} = 0.734$.

Due to missing data, we do not calculate $M(\mu, \nu^r)$ and $S(\mu, \nu^r)$.

Asia and the Pacific

Country	Government Response Rank	Vulnerability Rank	Prevalence Rank
Australia	1	27	25
New Zealand	2	28	26
Philippines	3	8	12
Indonesia	4	14	17
Thailand	5	12	11
Vietnam	6	20	18
India	7	10	15
Bangladesh	8	15	19
Nepal	9	17	16
Malaysia	10	21	14
Taiwan, China	11	24	27
Cambodia	12	5	4
Sri Lanka	13	19	22
Japan	14	25	28
Lao, PDR	15	9	10
Singapore	16	26	20
Myanmar	17	4	7
Mongolia	18	18	6
Timor-Leste	19	11	13
Korea (South)	20	22	23
China	21	13	21
Hong Kong, China	22	23	24
Brunei Darussalam	23	16	8
Papua New Guinea	24	7	9
Pakistan	25	2	3
Iran, Islamic Rep. of	26	6	5
Korea (North)	27	3	1
Afghanistan		1	2

Similarity Measure

Let ν denote the ranking with respect to vulnerability and π the ranking with respect to prevalence. Then

We find that $M(\nu, \pi) = \frac{361}{451} = 0.800$. The smallest possible similarity measure for 28 countries with no ties in the ranking can be obtained by substituting 28 for n in the formula $\frac{n+2}{3n+2}$. One obtains $30/86 = 0.349$. We have that $\frac{M(\nu,\pi)-0.349}{1-0.349} = \frac{0.800-0.349}{1-0.349} = 0.693$.

We also have that $S(\nu, \pi) = 1 - \frac{90}{792} = 0.886$. The smallest possible similarity measure for 28 countries with no ties can be obtained by substituting 28 for n in the formula $\frac{\frac{n}{2}+1}{n+1}$. One obtains $\frac{15}{29} = 0.517$. We have that $\frac{S(\nu,\pi)-0.517}{1-0.517} = \frac{0.886-0.517}{1-0.517} = 0.764$.

Let μ denote the ranking with respect to government response and ν^r the reverse ranking of the ranking with respect to vulnerability. Then $M(\mu, \nu^r) = \frac{313}{499} = 0.627$. The smallest possible similarity measure for 28 countries with no ties in the ranking is $\frac{n+2}{3n+2} = \frac{30}{86} = 0.349$. We have that $\frac{M(\mu,\nu^r)-0.349}{1-0.349} = \frac{0.627-0.349}{1-0.349} = 0.427$.

We also have that $S(\mu, \nu^r) = 1 - \frac{186}{812} = 1 - 0.229 = 0.771$. The smallest possible similarity measure for 28 countries with no ties is $\frac{n\backslash 2+1}{n+1} = \frac{15}{29} = 0.517$. We have that $\frac{S(\mu,\nu^r)-0.517}{1-0.517} = \frac{0.771-0.517}{1-0.517} = 0.526$.

Europe and Central Asia

Country	Government Response Rank	Vulnerability Rank	Prevalence Rank
Netherlands	1	44	43
United Kingdom	2	38	38
Sweden	3	46	49
Belgium	4	36	40
Croatia	5	23	8
Spain	6	37	35
Norway	7	45	42
Portugual	8	41	33
Montenegro	9	20	9
Cyprus	10	33	20
Macedonia	11	10	3
Austria	12	48	44
Georgia	13	17	18
Italy	14	26	34
Serbia	15	21	30
France	16	35	41
Latvia	17	29	24
Switzerland	18	49	45
Albania	19	11	5
Slovenia	20	32	36
Lithuania	21	28	10
Denmark	22	50	48
Hungary	23	31	26
Finland	24	43	47
Ireland	25	40	46
Germany	26	39	39
Bulgaria	27	24	16
Moldova, Rep. of	28	16	12
Greece	29	18	4
Kosovo	30	12	23
Poland	31	30	29
Armenia	32	13	13
Slovakia	33	27	31
Ukraine	34	3	7
Czech Republic	35	34	32
Isreal	36	19	25
Estonia	37	25	27
Bosnia and Herzegovina	38	9	28
Azerbaijan	39	6	17
Turkey	40	5	6
Iceland	41	47	37
Luxembourg	42	42	50
Romania	43	22	19
Kyrgyzstan	44	15	22
Belarus	45	8	2
Tajikistan	46	2	15
Kazakhstan	47	14	21
Uzbekenistan	48	7	14
Turkmenistan	49	1	1
Russia	50	4	11

Similarity Measure

Let ν denote the ranking with respect to vulnerability and π the ranking with respect to prevalence. Then

We find that $M(\nu, \pi) = \frac{1123}{1427} = 0.787$. The smallest possible similarity measure for 50 countries with no ties in the ranking can be obtained by substituting 50 for n in the formula $\frac{n+2}{3n+2}$. One obtains $52/152 = 0.342$. We have that $\frac{M(\nu,\pi)-0.342}{1-0.342} = \frac{0.787-0.342}{1-0.342} = 0.676$.

We also have that $S(\nu, \pi) = 1 - \frac{304}{2550} = 0.881$. The smallest possible similarity measure for 50 countries with no ties can be obtained by substituting 50 for n in the formula $\frac{\frac{n}{2}+1}{n+1}$. One obtains $\frac{26}{51} = 0.510$. We have that $\frac{S(\nu,\pi)-0.517}{1-0.517} = \frac{0.881-0.510}{1-0.510} = 0.758$.

Let μ denote the ranking with respect to government response and ν^r the reverse ranking of the ranking with respect to vulnerability. Then $M(\mu, \nu^r) = \frac{1017}{1533} = 0.663$. The smallest possible similarity measure for 50 countries with no ties in the ranking is $\frac{n+2}{3n+2} = \frac{52}{152} = 0.342$. We have that $\frac{M(\mu,\nu^r)-0.342}{1-0.342} = \frac{0.663-0.342}{1-0.342} = 0.488$.

We also have that $S(\mu, \nu^r) = 1 - \frac{516}{2550} = 1 - 0.202 = 0.798$. The smallest possible similarity measure for 50 countries with no ties is $\frac{n/2+1}{n+1} = \frac{26}{51} = 0.51$. We have that $\frac{S(\mu,\nu^r)-0.51}{1-0.51} = \frac{0.798-0.51}{1-0.51} = 0.588$.

6.2 Indices

In this section, we develop several indices to measure the susceptibility of a path with respect to trafficking in persons. We use the global scores of a country's government response and vulnerability provided in [60] and [59]. We normalize these scores. The formula used to do the normalization and the normalized scores are presented in the Appendix. Our work in this section is based on [38, 127].

We develop indices that can be used to measure the susceptibility of a route to human trafficking. The susceptibility is based on data providing a measure of a country's vulnerability for and its government response to human trafficking, [59]. We examine four routes through the Americas to the United States.

Let $G = (V, E)$ be a graph and let μ be a fuzzy subset of E.

Definition 6.2.1 Let P denote the path, $x_1x_2, \ldots, x_{n-1}x_n$. Define $\mu I(P) = \sum_{i=1}^{n-1} m_i\mu(x_ix_{i+1})$, where m_i is the number of subpaths in which $m_i(x_ix_{i+1})$ is the minimum weight in the subpath, $i = 1, 2, \ldots, n-1$. Then $\mu I(P)$ is called **μ-involvement** of P. As we will see, the μ-involvement of P can be used to measure the nonsusceptibility of P to trafficking since μ is a measure of two country's measure of government response and in other measures, the μ-value appears in the denominator.

It follows easily that $\sum_{i=1}^{n-1} m_i$ is the number of subpaths of P and so $\sum_{i=1}^{n-1} m_i = \frac{n(n-1)}{2}$.

Definition 6.2.2 Define $\mu AI(P) = \frac{2}{n(n-1)}\mu I(P)$. Then $\mu AI(P)$ is called the **average μ-involvement** of P.

The following index was influenced by the definition of a particular index given in [38]. We extend the definition to other types of fuzzy graphs. Let (σ, μ) be a fuzzy subgraph of a graph $G = (\sigma, \mu)$ and let (τ, ν) be a complementary fuzzy subgraph of G.

Definition 6.2.3 ([38]) Define the fuzzy subset δ of $V \times V$ by for all $(u, v) \in V \times V$

$$\delta(u, v) = \wedge\left\{\sum_{i=1}^{n}\left(\frac{1}{\mu(x_{i-1}x_i)} + \frac{1}{1-\nu(x_{i-1}x_i)}\right) | P_n \in \mathcal{P}_n(u, v), n \in \mathbb{N}\right\}.$$

Definition 6.2.4 ([38]) Define $e : V \to R$ by for all $u \in V, e(u) = \vee\{\delta(u, v)|v \in V\}$. The function e is called the **eccentricity** of u.

Let $n \in N$. Define $\mu^n(u, v) = \vee\{\mu(x_0x_1) \wedge ... \wedge \mu(x_{n-1}x_n)|P : u = x_0, x_1, \ldots, x_n = v$ is a path$\}$. Define $\nu^n(u, v) = \wedge\{\mu(x_0x_1) \vee ... \vee \mu(x_{n-1}x_n)|P : u = x_0, x_1, \ldots, x_n = v$ is a path$\}$. Let $\mu^\infty(u.v) = \vee\{\mu^n(u, v)|n \in \mathbb{N}\}$ and $\nu^\infty(u.v) = \wedge\{\mu^n(u, v)|n \in \mathbb{N}\}$. We say that (u, v) is a **strong pair** if $\mu(uv) \geq \mu^\infty(u, v)$ and $\nu(uv) \leq \nu^\infty(u, v)$. If (u, v) is a strong pair, then $\mu(uv) > 0$ and $\nu(uv) < 1$.

Theorem 6.2.5 ([38]) *Suppose (u, v) is a strong pair. Then* $\delta(u, v) = \frac{1}{\mu(uv)} + \frac{1}{1-\nu(uv)}$.

Let (σ, μ) be a fuzzy subgraph of a graph $G = (V, E)$ and let (τ, ν) be a complementary fuzzy subgraph of G. Then the pair (u, v) is called **μ-effective** if $\mu(uv) = \sigma(u) \wedge \sigma(v)$. The pair is called **$\nu$-effective** if $\nu(uv) = \tau(u) \vee \tau(v)$. The fuzzy subgraph $F = (\sigma, \mu)$ is called **complete** if (u, v) is μ-effective for all $u, v \in V$. The complementary fuzzy graph $C = (\tau, \nu)$ is called **complete** if (u, v) is ν-effective for all $u, v \in V$. μ is called **strong** if $\mu(uv) = \sigma(u) \wedge \sigma(v)$ for all $u, v \in V$ and ν is called **strong** if $\nu(uv) = \tau(u) \vee \tau(v)$ for all $u, v \in V$.

Lemma 6.2.6 ([38]) *Let (σ, μ) be a fuzzy subgraph of a graph $G = (V, E)$ and let (τ, ν) be a complementary fuzzy subgraph of G. Then*
(1) If (u, v) is μ-effective, then (u, v) is a μ-strong pair;
(2) If (u, v) is ν-effective, then (u, v) is a ν-strong pair.

Proposition 6.2.7 ([127]) *Let (σ, μ) be a fuzzy subgraph of a graph $G = (V, E)$ and let (τ, ν) be a complementary fuzzy subgraph of G. Suppose for all $x.y.u, v \in V$ that $\mu(x, y) \leq \mu(u, v)$ of and only if $\nu(x, y) \geq \mu(u, v)$. If (F, C) is complete, then for all $u \in V$,*

$$e(u) = \frac{1}{\wedge\{\sigma(v)|v \in V\}} + \frac{1}{1 - \vee\{\tau(v)|v \in V\}}.$$

Proof Since (F, C) is complete, we have for all $v \in V$, $v \neq u$, that $\mu(uv) = \sigma(u) \wedge \sigma(v)$ and $\nu(uv) = \tau(u) \vee \tau(v)$. Thus

$$\begin{aligned} e(u) &= \vee\delta(u, v)|v \in V\} \\ &= \vee\left\{\frac{1}{\mu(uv)} + \frac{1}{1 - \nu(uv)}|v \in V\right\} \\ &= \frac{1}{\wedge\{\sigma(v)|v \in V\}} + \frac{1}{1 - \vee\{\tau(v)|v \in V\}}, \end{aligned}$$

where the last equality holds by the hypothesis.

Definition 6.2.8 Let (σ, μ) be a fuzzy subgraph of a graph $G = (V, E)$ and let (τ, ν) be a complementary fuzzy subgraph of G. Let $V = \{x_1, \ldots, x_m\}$. Then $\frac{1}{\sigma(x_i)} + \frac{1}{1-\tau(x_i)}$ is called the **susceptibility** of x_i with respect to human trafficking. Let P : $x_1x_2, \ldots, x_{n-1}x_n$ be a path in G. Define $CRV(P) = \sum_{i-1}^{n}(\frac{1}{\sigma(x_i)} + \frac{1}{1-\tau(x_i)})$ is called the **susceptibility** of P with respect to human trafficking.

We use the length of a path defined by Rosenfeld to provide measures determining the success in combating human trafficking with respect to human response. The definition of the μ-length of a path $P : x_0, x_1, \ldots, x_n$ is $\sum_{i=1}^{n} \frac{1}{\mu(x_{i-1}x_i)}$. We let $\mu l(P)$ denote the μ-length of P.

The scores in [59], p. 194 are high if the government response is high. The scores in [59], p. 156 are high if the vulnerability of a country is high. These scores are normalized by a method to be discussed. The standard complement of these normalized scores are then high if the vulnerability is low. Consequently, we are more interested in the complement of the vulnerability scores. We use ν^c to denote the standard complement of ν. Then $\nu^c l(P) = \sum_{i=1}^{n} \frac{1}{\nu^c(x_{i-1}x_i)} = \sum_{i=1}^{n} \frac{1}{1-\nu(x_{i-1}x_i)}$. Note $(\nu^c)^c l(P) = \sum_{i=1}^{n} \frac{1}{1-\nu^c(x_{i-1}x_i)} = \sum_{i=1}^{n} \frac{1}{\nu(x_{i-1}x_i)} = \nu l(P)$.

It is important to note that the larger $\mu(uv)$ is, the smaller $\frac{1}{\mu(uv)}$ is and similarly for $\nu^c(uv)$. Thus in what follows, the larger the government response value is and the smaller the vulnerability value is, the smaller the susceptibility of trafficking for the path will be. The same sort of reasoning applies to δ and e defined above. Note also that the smallest $\frac{1}{\mu(uv)}$ can be is when $\mu(uv) = 1$. This is the case if and only if $\sigma(u) = \sigma(v) = 1$ when $\mu(uv) = i(\sigma(u), \sigma(v))$, where i is a t-norm. Also $\frac{1}{1-\nu(uv)}$ is the smallest when $\nu(uv) = 0$. This is the case if and only if $\tau(u) = \tau(v) = 0$ when $\nu(uv) = j(\tau(u), \tau(v))$, where j is a t-conorm. $\mu(uv)$ can be thought as the combined success in combatting human trafficking by means of government response of u and v and $\nu(uv)$ as the combined vulnerability.

Let $F = (\sigma, \mu)$ be a fuzzy subgraph of the $G = (V, E)$. Let $\mathcal{P}(u, v) = \{P|P$ is a path from u to $v\}$. We define the **Wiener Index** of F, written μWI, to be $\sum_{u,v \in V} l(u, v)$, where $l(u, v) = \wedge\{\mu l(P)|P \in \mathcal{P}(u, v)\}$. Then μWI provides a measure of susceptibility of trafficking with respect to a fuzzy graph. We let μAWI denote $\frac{1}{n}\mu WI$, where n is the number of edges in the fuzzy graph. The indices νWI and ν^c are defined similarly. It is important to note that the higher the success in combatting trafficking with respect to a path, the lower the Wiener Indices.

Suppose that we define $\mu(uv) = \sigma(u) \wedge \sigma(v)$. If for example, $\sigma(u) = 0.2$, then $\mu(uv) = 0.2$ for any value $s(v) \geq 0.2$. This does not seem realistic. Consequently, we use the t-norm product and define $\mu(uv) = \sigma(u)\sigma(v)$. By similar reasoning, we use the t-conorm algebraic sum and define $\nu(uv) = \tau(u) + \tau(v) - \tau(u)\tau(v)$.

We now apply our results to measure the susceptibility of the main routes from the Americas to the United States. The routes are determined from [163] and [127]. We first consider the following route:

Columbia (Co) $\to$ Costa Rica (CR) $\to$ El Salvador (ES) $\to$ Guatemala (Gu) $\to$ Mexico (Me) $\to$ U.S.

	Co		CR		ES		Gu		Me		U.S.
σ	0.530		0.662		0.480		0.587		0.686		0.951
μ		0.351		0.318		0.282		0.402		0.652	
τ	0.511		0.287		0.502		0.516		0.569		0.151
ν		0.655		0.645		0.759		0.791		0.631	

$$\begin{aligned}
\delta(\text{Columbia,U.S.}) &= \frac{1}{0.351} + \frac{1}{0.318} + \frac{1}{0.282} + \frac{1}{0.402} + \frac{1}{0.652} \\
&\quad + \frac{1}{1-0.655} + \frac{1}{1-0.645} + \frac{1}{1-0.758} + \frac{1}{1-0.791} + \frac{1}{1-0.631} \\
&= 2.85 + 3.14 + 3.57 + 2.49 + 1.53 + 2.90 + 2.82 + 4.13 + 4.78 + 2.71 \\
&= 30.92 = e(\text{Columbia}).
\end{aligned}$$

$$\begin{aligned}
CRV(\text{Columbia, U.S}) &= \frac{1}{0.530} + \frac{1}{0.662} + \frac{1}{0.480} + \frac{1}{0.587} + \frac{1}{0.686} + \frac{1}{0.951} \\
&= + \frac{1}{1-0.511} + \frac{1}{1-0.287} + \frac{1}{1-0.502} + \frac{1}{1-0.516} + \frac{1}{1-569} + \frac{1}{1-0.151} \\
&= 1.887 + 1.511 + 2.083 + 1.704 + 1.458 + 1.052 \\
&\quad +2.045 + 1.403 + 2.008 + 2.066 + 2.320 + 1.178 \\
&= 20.715.
\end{aligned}$$

This path from Columbia to U.S. consists of five edges. The edges Columbia$\to$ Costa Rica and Mexico $\to$ U.S. appear 5 times each and the other edges appear 8 times in subpaths. Thus

$$\begin{aligned}
\mu WI &= 5(2.85) + 8(3.14) + 8(3.57) + 8(2.49) + 5(1.53) \\
&= 14.25 + 25.12 + 28.56 + 19.92 + 7.65 = 95.5 \\
\mu AWI &= \frac{1}{5}(95.5) = 19.5 \\
\nu^c WI &= 5(2.90) + 8(2.82) + 8(4.13) + 8(4.78) + 5(2.71)
\end{aligned}$$

$$
\begin{aligned}
&= 14.50 + 22.56 + 33.04 + 38.24 + 13.55 = 121.89 \\
\nu^c AWI &= \frac{1}{5}(121.89) = 24.38 \\
\mu I &= 0.351 + 2(0.318) + 9(0.282) + 2(0.402) + 0.652 \\
&= 0.351 + 0.636 + 2.538 + 0.804 + 0.652 = 4.981 \\
\mu AI &= \frac{1}{15}(2.981) = 0.199 \\
\nu^c I &= 2(0.345) + 0.355 + 3(0.241) + 8(0.209) + 0.366. \\
&= .0.690 + 0.355 + 0.723 + 0.1.672 + 0.366 = 3.806 \\
\nu AI &= \frac{1}{15}(3.806) = 0.254
\end{aligned}
$$

We next consider the path Columbia (Co) → Ecuador (Ec) → Honduras (Ho) → Guatemala (Gu) → Mexico (Me) → U. S.

	Co		Ec		Ho		Gu		Me		U.S.
σ	0.530		0.603		0.474		0.587		0.686		0.951
μ		0.320		0.289		0.278		0.402		0.652	
τ	0.511		0.497		0.551		0.516		0.569		0.151
ν		0.754		0.774		0.783		0.791		0.634	

$$
\begin{aligned}
\delta(\text{Columbia,U.S.}) &= \frac{1}{0.320} + \frac{1}{0.289} + \frac{1}{0.278} + \frac{1}{0.402} + \frac{1}{0.652} \\
&\quad + \frac{1}{1-0.754} + \frac{1}{1-0.774} + \frac{1}{1-0.783} + \frac{1}{1-0.791} + \frac{1}{1-0.634} \\
&= 3.125 + 3.460 + 3.597 + 2.488 + 1.534 \\
&\quad + 1.065 + 4.425 + 4.608 + 4.785 + 2.732 \\
&= 34.819 = e(\text{Columbia})
\end{aligned}
$$

$$
\begin{aligned}
CRV(\text{Columbia, U.S}) &= \frac{1}{0.530} + \frac{1}{0.603} + \frac{1}{0.474} + \frac{1}{0.587} + \frac{1}{0.686} + \frac{1}{0.951} \\
&= + \frac{1}{1-0.511} + \frac{1}{1-0.497} + \frac{1}{1-0.551} + \frac{1}{1-0.516} + \frac{1}{1-569} + \frac{1}{1-0.151} \\
&= 1.887 + 1.658 + 2.110 + 1.704 + 1.458 + 1.052 \\
&\quad + 2.045 + 1.988 + 2.227 + 2.066 + 2.320 + 1.178 \\
&= 21.693
\end{aligned}
$$

This path from Columbia to U.S. consists of five edges. The edges Columbia→ Ecuador and Mexico→U.S. appear 5 times each and the other edges appear 8 times in subpaths. Thus

$$
\begin{aligned}
\mu WI &= 5(3.125) + 8(3.460) + 8(3.597) + 8(2.488) + 5(1.534) \\
&= 15.625 + 27.680 + 28.776 + 19.904 + 7.67 = 99.655 \\
\mu AWI &= \frac{1}{5}(99.655) = 19.931 \\
\nu^c WI &= 5(1.065) + 8(4.425) + 8(4.608) + 8(4.785) + 5(2.732) \\
&= 5.325 + 35.400 + 36.864 + 38.280 + 13.660 = 129.529 \\
\nu^c AWI &= \frac{1}{5}(129.529) = 25.906 \\
\mu I &= 0.320 + 2(0.289) + 9(0.278) + 2(0.402) + 0.652 \\
&= 0.320 + 0.578 + 2.502 + 0.804 + 0.652 = 4.856 \\
\mu AI &= \frac{1}{15}(4.856) = 0.324 \\
\nu^c I &= 0.246 + 2(0.226) + 3(0.217) + 8(0.209) + 0.366 \\
&= 0.246 + 0.452 + 0.651 + 0.1.672 + 0.366 = 3.387 \\
\nu^c AI &= \frac{1}{15}(3.387) = 0.229
\end{aligned}
$$

We next consider the path Nicaragua (Ni) → Costa Rica (CR) → El Salvador (ES) → Guatemala (Gu) → Mexico (Me) → U. S.

	Ni		CR		ES		Gu		Me		U.S.
σ	0.602		0.662		0.480		0.587		0.686		0.951
μ		0.399		0.318		0.282		0.403		0.652	
τ	0.433		0.287		0.502		0.516		0.569		0.151
ν		0.596		0.645		0.759		0.791		0.634	

$$
\begin{aligned}
\delta(\text{Nicaragua, U.S.}) &= \frac{1}{0.399} + \frac{1}{0.318} + \frac{1}{0.282} + \frac{1}{0.403} + \frac{1}{0.652} \\
&\quad + \frac{1}{1-0.596} + \frac{1}{1-0.645} + \frac{1}{1-0.759} + \frac{1}{1-0.791} + \frac{1}{1-0.634} \\
&= 2.506 + 3.144 + 3.546 + 2.481 + 1.534 \\
&\quad + 2.475 + 2.817 + 4.491 + 4.785 + 2.732 \\
&= 30.511 = e(\text{Nicaragua}) \\
CRV(\text{Nicaragua, U.S.}) &= \frac{1}{0.602} + \frac{1}{0.662} + \frac{1}{0.480} + \frac{1}{0.587} + \frac{1}{0.686} + \frac{1}{0.951} \\
&\quad + \frac{1}{1.0.433} + \frac{1}{1-0.287} + \frac{1}{1-0.502} + \frac{1}{1-0.516} + \frac{1}{1-0.569} + \frac{1}{1-0.151} \\
&= 1.661 + 1.510 + 2.083 + 1.703 + 1.457 + 1.051 \\
&\quad + 1.763 + 1.402 + 2.008 + 2.066 + 2.320 + 1.177 \\
&= 20.201
\end{aligned}
$$

This path from Nicaragua to U.S. consists of five edges. The edges Nicaragua→ Costa Rica and Mexico→U.S. appear 5 times each and the other edges appear 8 times in subpaths. Thus

$$
\begin{aligned}
\mu WI &= 5(2.506) + 8(3.144) + 8(3.546) + (2.481) + 5(1.534) \\
&= 12.530 + 25.152 + 28.368 + 19.848 + 7.670 = 93.568 \\
\mu AWI &= \frac{1}{5}(93.568) = 18.714 \\
\nu^c WI &= 5(2.475) + 8(2.817) + 8(4.491) + 8(4.785) + 5(2.732) \\
&= 12.375 + 22.536 + 36.928 + 38.280 + 13.660 = 123.779 \\
\nu^c AWI &= \frac{1}{5}(123.779) = 24.756 \\
\mu I &= 0.300 + 2(0.318) + 9(0.282) + 2(0.403) + 0.6525 = 5.031 \\
\mu AI &= \frac{1}{15}(5.031) = 0.335 \\
\nu^c I &= 0.404 + 2(0.355) + 3(0.241) + 8(0.209) + 0.366 \\
&= 0.404 + 0.710 + 0.723 + 1.672 + 0.366 + 3.875 = 3.875 \\
\nu^c AI &= \frac{1}{15}(3.875) = 0.258
\end{aligned}
$$

We next consider the path Dominican Republic (DR) $\rightarrow$ Panama (Pa) $\rightarrow$ Costa Rica (CR) $\rightarrow$ El Salvador (ES) $\rightarrow$ Guatemala (Gu) $\rightarrow$ Mexico (Me) $\rightarrow$ U. S.

	DR		Pa		CR		ES		Gu		Me		U.S.
σ	0.763		0.569		0.662		0.480		0.587		0.686		0.951
μ		0.434		0.377		0.318		0.282		0.403		0.652	
τ	0.425		0.358		0.287		0.502		0.516		0.569		0.151
ν		0.631		0.542		0.645		0.759		0.791		0.634	

$$
\begin{aligned}
\delta(\text{Dominican Rep., U.S}) &= \frac{1}{0.434} + \frac{1}{0.377} + \frac{1}{0.318} + \frac{1}{0.282} + \frac{1}{0.403} + \frac{1}{0.652} \\
&\quad + \frac{1}{1-0.631} + \frac{1}{1-0.542} + \frac{1}{1-0.645} + \frac{1}{1-0.759} \\
&\quad + \frac{1}{1-0.791} + \frac{1}{1-0.634} \\
&= 2.304 + 2.652 + 3.144 + 3.546 + 2.481 + 1.534 \\
&\quad + 2.710 + 2.183 + 2.816 + 4.149 + 4.784 + 2.732 \\
&= 35.035 \\
CRV(\text{Dominican Rep., U.S.}) &= \frac{1}{0.763} + \frac{1}{0.569} + \frac{1}{0.602} + \frac{1}{0.480} + \frac{1}{0.587} + \frac{1}{0.686} + \frac{1}{0.951} \\
&\quad + \frac{1}{1-0.425} + \frac{1}{1-0.358} + \frac{1}{1-0-.287} + \frac{1}{1-0.502} + \frac{1}{1-0.516} \\
&\quad + \frac{1}{1-0.569} + \frac{1}{1-0.151} \\
&= 1.310 + 1.757 + 1.510 + 2.083 + 1.703 + 1.457 + 1.051 \\
&\quad + 1.739 + 1.557 + 1.402 + 2.008 + 2.066 + 2.320 + 1.177 \\
&= 23.140
\end{aligned}
$$

The path from the Dominican Republic to U.S. consists of 6 edges. The edges Dominican Republic to Panama and Mexico to U.S. appear 6 times each, the edges Panama to Costa Rica and Guatemala to Mexico appear 10 times each, and the edges Costa Rica to El Salvador and El Salvador to Guatemala appear 12 times in subpaths. Thus

$$\begin{aligned}
\mu WI &= 6(2.304) + 10(2.652) + 12(3.144) + 12(3.546) + 10(2.481) + 6(1.534 \\
&= 13.824 + 26.52 + 37.728 + 42.552 + 24.81 + 9.204 = 154.638 \\
\mu AWI &= \frac{1}{6}(154.638) = 25.773 \\
\nu^c WI &= 6(2.710) + 10(2.183) + 12(2.816) + 12(4.149) + 10(4.784) + 6(2.732) \\
&= 16.26 + 21.83 + 33.792 + 49.788 + 47.84 + 16.392 = 185.902 \\
\nu^c AWI &= \frac{1}{6}(185.902) = 30.984 \\
\mu I &= 0.434 + 2(0.377) + 3(0.318) + 12(0.282) + 2(0.403) + 0.652 = 6.984 \\
\mu AI &= \frac{1}{21}(6.984) = 0.333 \\
\nu^c I &= 2(0.369) + 0.458 + 3(0.355) + 4(0.241) + 10(0.209) + 0.366 \\
&= 0.739 + 0.458 + 1.065 + 0.964 + 2.090 + 0.366 = 5.681 \\
\nu^c AI &= \frac{1}{21}(5.681) = 0.271
\end{aligned}$$

Appendix 6.1

Let X be a finite set of positive real numbers. We use the following formula to normalize the data given [59]. Define $n : X \to [0, 1]$ by for all $x \in X$, $n(x) = \frac{x - \min X}{\max X - \min X}$, where $\min X$ is the smallest element in X and $\max X$ is the largest element in X. This yields the following table.

Country	Government Response	Vulnerability
Columbia	0.530	0.511
Costa Rica	0.662	0.287
Dominican Republic	0.763	0.425
Ecuador	0.603	0.407
El Salvador	0.480	0.502
Guatemala	0.587	0.516
Honduras	0.474	0.551
Mexico	0.686	0.569
Nicaragua	0.602	0.433
Panama	0.569	0.358
United States	0.951	0.151

Chapter 7
Global Starvation and Coronavirus

According to the New York Times, experts say that the world has never faced a hunger emergency like this. The number of people facing acute hunger could double to 265 million by the end of the year. The coronavirus pandemic has brought hunger to millions of people around the world. National lockdowns and social distancing measures are drying up work and incomes, and are likely to disrupt agricultural production and supply routes leaving millions to worry how they will get enough to eat. Already 135 million people had been facing acute food shortages, but now with the pandemic, 130 million more could go hungry in 2020, said Arif Husain, chief economist at the World Food Program, a United Nations agency. Altogether, 365 million people could be pushed to the brink of starvation by year's end.

This hunger crisis, experts say, is global and caused by a multitude of factors linked to the coronavirus pandemic and the ensuing interruption of the economic order: The sudden loss of income for countless millions who were living hand-to-mouth; the collapse in oil prices; widespread shortages of hard currency from tourism drying up; overseas workers not having enough earnings to send home; and ongoing problems like climate change, violence, population dislocations and humanitarian disasters.

7.1 Global Food Security

The following discussion is from [56]. Food security is defined as the state in which people at all times have physical, social and economic access to sufficient and nutritious food that meet their dietary needs and food preferences for a healthy and active life.

Using this definition adapted from the 1996 World Food Summit, the Global Food Security Index considers the core issues of affordability, availability, and quality across a set of 113 countries. The index is a dynamic quantitative and qualitative

J. N. Mordeson and S. Mathew, *Mathematics of Uncertainty for Coping with World Challenges*, Studies in Systems, Decision and Control 353,
https://doi.org/10.1007/978-3-030-68684-0_7

scoring model constructed from 34 unique indicators, that measure these drivers of food security across both developing and developed countries. The overall goal of the study is to assess which countries are most and least vulnerable to food insecurity through the categories of Affordability, Availability, and Quality and Safety. The index also looks at the impact that Natural Resources and Resilience have on food security.

While food security research is the subject of many organizations worldwide, the effort is distinct for a number of reasons. This index is the first to examine food security comprehensively across the three internationally established dimensions. Moreover, the study looks beyond hunger to the underlying factors effecting food insecurity. A number of unique qualitative indicators was created,. Many of the indicators relate to government policy, to capture drivers of food security which are not currently measured in any international data set.

The methodology for The global Food Security Index was developed by The Economist Intelligence Unit with consultation from a peer of experts. For a full overview of the GFSI methodology, download the report and consult the Methodology section. in the Appendix. The peer panel meeting was designed to engage a panel of experts from the academic, non-profit, and government sectors to help select and prioritize food security indicators. The goal of the peer panel meeting was to review the framework, selection of indicators, weighting and overall construction of the Global Food Security Index.

F_1 : Affordability: Measures the ability of consumers to purchase food, their vulnerability to price shocks and presence of programs and policies to support customers when shocks occur.

F_2 : Availability: Measures the sufficiency of the national food supply, the risk of supply disruption, national capacity to disseminate food research efforts to expand agricultural output.

F_3 : Quality and Safety: Measures the variety and nutritional quality of average diets, as well as the safety of food.

Natural Resources and Adjustment: Assesses a country's exposure to the impacts of climate change; its susceptibility to natural resources risk, and how the country is adapting to these risks.

The following parameters are considered for giving ranking to the countries, [56]: nutritional standards, urban absorption capacity, food consumption as a share of household expenditure, food loss, protein equality, agricultural import tariffs, diet diversification, agriculture infrastructure, volatility of agricultural production, proportion of population under global poverty line, gross domestic product per capita (USS$ PPP), presence of food safety net programmes, access to financing farmers, public expenditure on agricultural R&D, corruption, sufficiency of supply, food safety.

The relative importance of the factors F_1, F_2, and F_3 were given by three experts. Their opinions were used to determine the Guiasu and Dempster weights as given in the following table. The equations then follow from the table.

	E_1	E_2	E_3	Row Avg.	Row Product
F_1	0.40	0.36	0.38	0.38	0.05472
F_2	0.41	0.44	0.41	0.42	0.07396
F_3	0.19	0.20	0.21	0.20	0.00798
Col. Sum					0.13666

$$G = 0.38F_1 + 0.42F_2 + 0.20F_3$$
$$D = 0.40F_1 + 0.54F_2 + 0.06F_3$$

These equations were used to determine the score and rankings provided in the next table. The values for F_1, F_2, and F_3 were taken from [56]. They are the same values used to determine the GFSI scores and ranking in [56]. The GFSI scores for the countries are given in [56]. See Table 7.1.

In the following, μ denotes the Guiasu ranking, ν denotes the Dempster ranking, and ρ denotes the GFSI ranking. We note that the three rankings have a very high degree of similarity.

$M(\mu, \nu) = \frac{6324}{6558} = 0.964$, $M(\mu, \rho) = \frac{6400.5}{6481.5} = 0.988$, $M(\nu, \rho) = \frac{6352.5}{6529.5} = 0.973$. $n = 113. \frac{n+1}{3n-1} = \frac{114}{338} = 0.337$.

$$\frac{M(\mu, \nu) - 0.337}{1 - 0.337} = \frac{0.964 - 0.337}{1 - 0.337} = \frac{0.627}{0.663} = 0.946$$
$$\frac{M(\mu, \rho) - 0.337}{1 - 0.337} = \frac{0.988 - 0.337}{1 - 0.337} = \frac{0.651}{0.663} = 0.982$$
$$\frac{M(\nu, \rho) - 0.337}{1 - 0.337} = \frac{0.973 - 0.337}{1 - 0.337} = \frac{0.636}{0.663} = 0.959$$

$S(\mu, \nu) = 1 - \frac{236}{12882} = 1 - 0.018 = 0.982$, $S(\mu, \rho) = 1 - \frac{78}{12882} = 1 - 0.006 = 0.994$, $S(\nu, \rho) = 1 - \frac{177}{12882} = 1 - 0.014 = 0.986$.

$n = 113. \frac{1}{2} + \frac{1}{2n} = \frac{1}{2} + \frac{1}{226} = 0.504$.

$$\frac{S(\mu, \nu) - 0.504}{1 - 0.504} = \frac{0.945 - 0.504}{1 - 0.504} = \frac{0.441}{0.496} = 0.889$$
$$\frac{S(\mu, \rho) - 0.504}{1 - 0.504} = \frac{0.994 - 0.504}{1 - 0.504} = \frac{0.490}{0.496} = 0.988$$
$$\frac{S(\nu, \rho) - 0.504}{1 - 0.504} = \frac{0.986 - 0.504}{1 - 0.504} = \frac{0.482}{0.496} = 0.972$$

Some key findings of the 2019 GFSI follow. Singapore and Ireland remain the two most secure countries, with Kuwait, Qatar and Malawi making the most improvements since 2018. The percentage of cultivated land for irrigation is inadequate to meet global needs. Overall agricultural infrastructure has improved in Qatar, Belarus, Slovakia, Australia, and Kuwait. Public expenditure on agriculture is stagnate. Global

Table 7.1 Global food security

Country	GFSI Rank	Score/Guiasu Rank	Score/Dempster Rank
Singapore	1	86.992/1	87.774 /1
Ireland	2	84.186/2	82.934 /3
United States	3	83.918/3	82.588/4
Switzerland	4	82.890/7	83.334/2
Finland	5	83.330/4	81.592 /7
Norway	6	83.242/5	81.930 /5
Sweden	7	82.982/6	81.538 /8
Canada	8	82.594/8	81.722 /6
Netherlands	9	82.312/9	80.722/12
Austria	10	81.684/10	81.470 /9
Germany	11	81.444/11	81.462 /10
Australia	12	81.270/14	81.068/11
Qatar	13	81.282/13	79.166 /15
Denmark	14	81.308/12	79.784 /14
Belgium	15	80.856/15	79.942/13
France	16	80.720/16	79.138 /16
United Kingdom	17	79.196/18	78.470/18
Isreal	18	79.212/17	77.972 /19
New Zealand	19	78.558/19	79.020/17
Portugal	20	78.272/20	76.086 /20
Japan	21	76.472/22	75.902 /21
UAE	22	76.578/21	75.028 /23
Italy	23	75.976/23	74.664 /24
Poland	24	75.824/25	74.632 /25
Chile	25	75.476/26	75.184 /22
Spain	26	75.892/24	73.588 /26
Kuwait	27	74.824/27	73.436 /28
Malaysia	28	73.600/30	73.474 /27
South Korea	29	73.688/29	73.256/29
Saudi Arabia	30	73.4540/31	72.302/31
Greece	31	74.002/28	71.326 /34
Czech Republic	32	72.854/32	72.928/30
Uruguay	33	72.808/33	72.136 /33
Hungary	34	72.566/34	72.244 /32
China	35	71.042/37	70.402 /35
Belarus	36	71.338/35	69.178 /37
Argentina	37	71.166/36	68.830/41
Romania	38	69.960/40	70.288 /36
Brazil	39	70.756/38	67.592 /46

(continued)

Table 7.1 (continued)

Country	GFSI Rank	Score/Guiasu Rank	Score/Dempster Rank
Costa Rica	40	70.350/39	68.850/40
Turkey	41	69.862/41	69.138 /38
Russia	42	69.746/42	68.574 /42
Columbia	43	69.418/44	69.062 /39
Mexico	43	69.668/43	68.114 /44
Panama	45	68.906/45	67.902 /45
Oman	46	68.636/46	66.688 /48
Slovakia	47	67.830/47	68.538 /43
Kazakhstan	48	67.344/48	66.256/50
South Africa	49	67.234/49	67.122/47
Bahrain	50	66.148/50	66.576 /49
Bulgaria	51	66.144/51	64.876 /55
Thailand	52	64.472/54	65.694 /51
Azerbaijan	53	64.278/55	65.328/53
Vietnam	54	63.952/57	65.380 /52
Egypt	55	64.552/52	64.902 /54
Dominican Republic	56	64.072/56	64.038/58
Botswana	57	63.476/58	64.298 /56
Peru	58	63.118/59	63.124 /59
Ghana	59	64.528/53	63.264 /60
Morocco	60	62.714/60	62.982 /61
Serbia	61	62.702/61	61.888 /62
Indonesia	62	61.918/62	64.088/57
Ecuador	63	61.614/63	61.558 /64
Jordan	64	60.646/64	61.044 /66
Philippines	65	60.476/67	61.736/63
Sri Lanka	66	60.380/68	61.544/65
El Salvador	67	60.636/65	60.698/67
Guatamala	68	60.506/66	60.674/68
Tunesia	69	60.170/69	59.652 /71
Algeria	70	59.458/70	60.072 /69
Uzbekistan	71	58.750/71	59.198/72
India	72	58.324/72	60.036 /70
Honduras	73	58.132/74	57.728 /74
Paraguay	74	58.248/73	55.620 /79
Bolivia	75	57.644/75	56.818 /76

(continued)

Table 7.1 (continued)

Country	GFSI Rank	Score/Guiasu Rank	Score/Dempster Rank
Ukraine	76	57.202/76	56.136 /78
Myanmar	77	56.742/77	57.606 /75
Pakistan	78	56.130/79	57.974 /73
Nepal	79	56.238/78	56.538 /77
Mali	80	52.864/82	54.408 /82
Senegal	81	54.390/80	54.300 /83
Nicaragua	82	53.888/81	54.164/84
Bangladesh	83	52.088/83	55.588/80
Cote d'Ivoirre	84	51.352/84	54.760/81
Benin	85	50.806/85	51.870 /85
Kenya	86	50.346/86	51.192 /88
Burkina Faso	87	49.658/88	51.482/86
Cameroon	88	49.798/87	50.004 /92
Niger	89	49.068/89	51.268 /87
Cambodia	90	48.668/92	50.730 /89
Ethiopia	91	48.778/91	50.624 /90
Laos	92	48.562/93	50.148 /91
Tajikstan	93	48.926/90	48.510/94
Nigeria	94	48.528/94	47.934 /97
Rwanda	95	48.184/95	48.510 /95
Tanzania	96	47.486/96	48.010 /96
Guinea	97	45.820/98	48.996 /93
Uganda	98	46.334/97	45.836 /99
Sudan	99	45.746/99	45.576 /101
Angola	100	45.484/100	45.084 /102
Zambia	101	43.898/101	46.114 /98
Togo	102	43.352/102	45.588 /100
Haiti	103	42.926/103	43.658 /105
Malawi	104	42.004/104	43.990 /104
Mozambique	105	40.388/105	44.102/103
Sierra Leone	106	38.550/107	39.918/107
Syria	107	38.766/106	37.630 /109
Madagascar	108	37.180/108	40.284/106
Chad	109	36.672/109	36.976 /110
Congo (Dem. Rep.)	110	34.934/111	37.708/108
Yemen	111	35.342/110	35.456 /111
Burundi	112	34.332/112	34.098/112
Venezuela	113	32.908/113	27.722/113

food prices are rising worldwide. Food security is deteriorating in countries beset by turmoil. These countries include Yemen, Syria, and Venezuela. Dependency on food aid is highest in Syria, Yemen, Benin, and Haiti. Please see [56] for a much fuller account (Table 7.1).

7.2 Global Hunger Index

The Global Hunger Index (GHI) presents a three-step process that results in GHI scores on a 100-point GHI Severity Scale, where 0 is the best score (no hunger) and 100 is the worst, [57]. The composition of the global hunger index is broken into three parts, Inadequate Food Supply, Child Mortality, and Child Undernutrition. Child Undernutrition is broken into two parts,Wasting and Stunting.

F_1Inadequate Food Supply: Undernourishment: Measures inadequate food supply, an important indicator of hunger; Refers to the entire population, both children and adults, Used as a lead indicator for international hunger targets, including the SDGs.

F_2 Child Mortality: Under-five mortality rate: Death is the most serious consequence of hunger, and children are the most vulnerable; Improves The GHI's ability to reflect micronutrient deficiencies; Wasting and stunting only partially capture the mortality risk undernutrition.

F_3 Child Undernutrition: Goes beyond calorie availability, considers aspects of diet quality and utilization; Children are particularly vulnerable to nutritional deficiencies; Is sensitive to uneven distribution of food within the household; Stunting and wasting are nutrition indicators for the SDGs.

F_{31} Wasting

F_{32} Stunting

The GHI rates $F_1 F_2$, and F_3 of equal importance. F_3 is composed of F_{31} and F_{32} which are considered to be of equal importance. This leads to the following equations, where F denotes the overall severity rating.

$$F = \frac{1}{3}F_1 + \frac{1}{3}F_2 + \frac{1}{3}F_3,$$
$$F_3 = \frac{1}{2}F_{31} + \frac{1}{2}F_{32}.$$

Thus

$$F = \frac{1}{3}F_1 + \frac{1}{3}F_2 + \frac{1}{6}F_{31} + \frac{1}{6}F_{32}.$$

From these equations, we can see how techniques previously used in the book and in [57] could be used for further examination of global hunger.

The following is from [57]. The indicators included in the GHI formula reflect caloric deficiencies as well as poor nutrition. The under nourishment indicator captures the hunger situation of the population as a whole, while the indicators specific to children reflect the nutrition status within a particularly vulnerable subset of the population for whom a lack of dietary energy, protein, and/or micronutrients (essential vitamins and minerals) leads to a high risk of illness, poor physical and cognitive development, and death. The inclusion of both child wasting and child stunting allows the GHI to document both acute and chronic undernutrition. By combining multiple indicators, the index reduces the effects of random measurement errors.

In [57], a detailed explanation of how the formulas are used is given. The rankings of 117 countries is also provided.

7.3 Emerging, Developed, and Frontier Markets

In this section, we consider the rankings of the countries in Sect. 3.2. The overall climate vulnerability scores can be found in Table 3.1. We rerank the countries with respect to the category assigned to them in [134] as developed market (DM), emerging market (EM), or frontier market (FM). We name this ranking the Fragile ranking due to the title of the paper, [134].

Emerging nations are growing rapidly and becoming more important in world economies, while developing nations are struggling and still need help from trade partners around the world.

A developed market is one of the nations that belong to the Organization for Economic Cooperation and Development (OECD). The participants agree to collect and share information about their economies in order to develop new policies for the themselves and to assist in the development of the rest of the world. A developed market is a nation that is usually thought of as fully developed and has demonstrated a commitment to democracy.

A frontier market is a type of developing country which is more developed than the least developing countries, but too small, risky, or illiquid to be generally considered an emerging market. Recall that the lower the score, the higher the climate vulnerability. See Tables 7.2 and 7.3.

In what follows, we let μ denote the ranking for the Guiasu method, ν the ranking for the Dempster method, and ρ the ranking for the Fragile method.

We have that $M(\mu, \nu) = \frac{307}{343} = 0.895$, $M(\mu, \rho) = \frac{300}{350} = 0.857$, $M(\nu, \rho) = \frac{284}{366}$ $= 0.776$.

$n = 25. \frac{n+1}{3n-1} = \frac{26}{74} = 0.352.$

$$\frac{M(\mu, \nu) - 0.352}{1 - 0.352} = \frac{0.895 - 0.352}{1 - 0.352} = \frac{0.543}{0.648} = 0.838$$
$$\frac{M(\mu, \rho) - 0.352}{1 - 0.352} = \frac{0.857 - 0.352}{1 - 0.352} = \frac{0.505}{0.648} = 0.779$$
$$\frac{M(\nu, \nu) - 0.352}{1 - 0.352} = \frac{0.776 - 0.352}{1 - 0.352} = \frac{0.424}{0.648} = 0.654$$

$S(\mu, \nu) = 1 - \frac{36}{650} = 1 - 0.055 = 0.945$, $S(\mu, \rho) = 1 - \frac{50}{650} = 1 - 0.077 = 0.923$, $S(\nu, \rho) = 1 - \frac{82}{650} = 1 - 0.126 = 0.874$.

$n = 25. \frac{1}{2} + \frac{1}{2n} = \frac{1}{2} + \frac{1}{50} = 0.520$.

$$\frac{S(\mu, \nu) - 0.520}{1 - 0.520} = \frac{0.945 - 0.520}{1 - 0.520} = \frac{0.425}{0.480} = 0.885$$
$$\frac{S(\mu, \rho) - 0.520}{1 - 0.520} = \frac{0.923 - 0.520}{1 - 0.520} = \frac{0.400}{0.480} = 0.833$$
$$\frac{S(\nu, \rho) - 0.520}{1 - 0.520} = \frac{0.874 - 0.520}{1 - 0.520} = \frac{0.354}{0.480} = 0.738$$

We have that $M(\mu, \nu) = \frac{198}{222} = 0.892$, $M(\mu, \rho) = \frac{198}{222} = 0.892$, $M(\nu, \rho) = \frac{189}{231} = 0.818$.

$n = 20. \frac{n+2}{3n+2} = \frac{22}{62} = 0.355$.

$$\frac{M(\mu, \nu) - 0.355}{1 - 0.355} = \frac{0.892 - 0.355}{1 - 0.355} = \frac{0.537}{0.645} = 0.833$$
$$\frac{M(\mu, \rho) - 0.355}{1 - 0.355} = \frac{0.892 - 0.355}{1 - 0.355} = \frac{0.537}{0.645} = 0.833$$
$$\frac{M(\nu, \rho) - 0.355}{1 - 0.355} = \frac{0.818 - 0.355}{1 - 0.355} = \frac{0.463}{0.645} = 0.718$$

$S(\mu, \nu) = 1 - \frac{24}{420} = 1 - 0.057 = 0.943$, $S(\mu, \rho) = 1 - \frac{24}{420} = 1 - 0.057 = 0.943$, $S(\nu, \rho) = 1 - \frac{42}{420} = 1 - 0.100 = 0.900$.

$n = 20. (\frac{n}{2} + 1)/(n + 1) = \frac{11}{21} = 0.524$.

$$\frac{S(\mu, \nu) - 0.524}{1 - 0.524} = \frac{0.943 - 0.524}{1 - 0.524} = \frac{0.419}{0.476} = 0.880$$
$$\frac{S(\mu, \rho) - 0.524}{1 - 0.524} = \frac{0.943 - 0.524}{1 - 0.524} = \frac{0.419}{0.476} = 0.880$$
$$\frac{S(\nu, \rho) - 0.524}{1 - 0.524} = \frac{0.900 - 0.524}{1 - 0.524} = \frac{0.376}{0.476} = 0.790$$

We have that $M(\mu, \nu) = \frac{241}{265} = 0.909$, $M(\mu, \rho) = \frac{245}{261} = 0.939$, $M(\nu, \rho) = \frac{235}{271} = 0.867$.

$n = 22. \frac{n+2}{3n+2} = \frac{24}{68} = 0.324$.

$$\frac{M(\mu, \nu) - 0.324}{1 - 0.324} = \frac{0.909 - 0.324}{1 - 0.324} = \frac{0.585}{0.676} = 0.865$$
$$\frac{M(\mu, \rho) - 0.324}{1 - 0.324} = \frac{0.939 - 0.324}{1 - 0.324} = \frac{0.615}{0.676} = 0.910$$
$$\frac{M(\nu, \rho) - 0.324}{1 - 0.324} = \frac{0.867 - 0.324}{1 - 0.324} = \frac{0.543}{0.676} = 0.803$$

$S(\mu, \nu) = 1 - \frac{24}{506}$=1 $- 0.047 = 0.953$, $S(\mu, \rho) = 1 - \frac{16}{506} = 1 - 0.032$=0.968, $S(\nu, \rho) = 1 - \frac{35}{506} = 1 - 0.069 = 0.931$.

$n = 22.(\frac{n}{2} + 1)/(n + 1) = \frac{12}{23} = 0.522$.

$$\frac{S(\mu, \nu) - 0.522}{1 - 0.522} = \frac{0.953 - 0.522}{1 - 0.522} = \frac{0.431}{0.478} = 0.902$$
$$\frac{S(\mu, \rho) - 0.522}{1 - 0.522} = \frac{0.968 - 0.522}{1 - 0.522} = \frac{0.446}{0.478} = 0.933$$
$$\frac{S(\nu, \rho) - 0.522}{1 - 0.522} = \frac{0.931 - 0.522}{1 - 0.522} = \frac{0.409}{0.478} = 0.856$$

Table 7.2 Climate vulnerability

EM

Country (Rank Fragile)	Rank Guiasu	Rank Dempster
India (1)	1	1
Pakistan (2)	4	8
Philippines (3)	3	6
Columbia (4)	2	2
Mexico (5)	6	5
Kenya (6)	7	12
South Africa (7)	5	3
Thailand (8)	9	10
Indonesia (9)	13	13
Egypt (10)	11	11
Brazil (11)	14	17
Malaysia (12)	10	7
Peru (13)	16	15
S.Arabia (14)	8	4
Greece (15)	15	14
China (16)	17	18
Poland (17)	20	20
Qatar (18)	12	9
Czech Rep. (19)	21	21
Russia (20)	19	19
Hungary (21)	23	24
Turkey (22)	22	22
UAE (23)	18	16
Chile (24)	25	25
S. Korea (25)	24	23

Table 7.3 Climate vulnerability

EM		
Country (Rank Fragile)	Rank Guiiasu	Rank Dempster
Bangladesh (1)	2	3
Oman (2)	1	1
Sri Lanka (3)	4	6
Lebanon (4)	3	2
Vietnam (5)	6	8
Nigeria (6)	5	7
Morocco (7)	7	9
Serbia (8)	11	12
Bahrain (9)	10	5
Tunesia (10)	9	10
Argentina (11)	14	14
Mauritius (12)	12	13
Kuwait (13)	8	4
Jordan (14)	15	15
Kazakhistan (15)	13	11
Romania (16)	17	17
Slovenia (17)	16	16
Croatia (18)	18	18
Lithuania (19)	19	19
Estonia (20)	20	20

We see that the three methods of ranking have a high degree of similarity for EM, FM, and DM (Tables 7.2 and 7.3).

7.4 Coronavirus

Data is readily available to most countries concerning the coronavirus. Citizens of these countries are knowledgeable concerning the spread of the virus and how to contain it. However, as reported by the Omaha World Herald, the rapid spread of the coronavirus has raised fears about the world's refugees and internationally displaced people, many of whom live in war-ravaged countries ill-equipped to test for the virus or contain an outbreak. Limiting testing means the virus may be able to spread unchecked until people show symptoms. More than 70 million people world-wide have been forced out of their homes, according to the U. N. refugees agency, although the vast majority do not live in camps (Table 7.4).

Table 7.4 Climate vulnerability

DM

Country (Rank Fragile)	Rank Guiasu	Rank Dempster
Isreal (1)	1	1
Austalia (2)	2	2
UK (3)	3	6
Portugal (4)	5	7
USA (5)	7	3
Belgium (6)	4	4
Japan (7)	6	5
France (8)	9	11
Italy (9)	10	10
Singapore (10)	8	8
Spain (11)	13	13
Germany (12)	12	12
Netherlands (13)	11	9
Austria (14)	14	15
Canada (15)	15	14
Switzerland (16)	16	18
Denmark (17)	17	17
Ireland (18)	18	19
New Zealand (19)	19	20
Norway (20)	20	16
Sweden (21)	22	22
Finland (22)	21	21

More than a million Rahingya Muslims from neighboring Myanmar are parked into the world's largest refugee camp in Bangladesh. Most fled a scorch-earth campaign by Myanmar's government and Buddhist mobs in 2017. Aid workers are rushing to build isolation wards.

The nearly 10-year civil war in Syria has displaced half of Syria's population of 23 million. The northwestern Iflab province, home to 4 million people, has just one small health facility equipped to receive suspected coronavirus cases. Authorities have conducted around 200 tests, all of which came back negative.

Iraq is home to some 1.4 million people displaced by the war against the Islamic State; many of them live in camps, but aid refugees have struggled to provide services amid restrictions imposed to prevent the virus spread.

Decades of war have displaced millions of Afghans, and most fled to neighboring Iran and Pakistan. Iran is battling the deadliest caronavirus outbreak in the Middle East. The International Organization of Migration says 200,000 Afghans have returned from Iran this year, as the outbreak there accelerated. that could potentially contribute to the spread of the virus in Afghanistan, which is still mired in conflict and poverty.

In South Sudan, more than 180,000 people live in crowded U. N.—run camps across the country after a five-year civil war ended in 2018. South Sudan has reported just four cases, with none inside the camps, but one was a woman who had been in the country for five weeks.

More than 2.2 million people live in displacement camps after being uprooted by cycles of drought and ever present threat of jihadis. Another 400,000 Somalis live in two sprawling refugee camps in neighboring Kenya. There is no testing in either camp nor are their ventilators.

More than 5 million people have been uprooted by conflict within Congo's borders, and more than half a million refugees have fled from neighboring countries. Violence in the east has displaced thousands, even as the region has grappled with an Ebola outbreak that has killed more than 2260 people.

Nearly 5 million Venezuelans have fled economic collapse, crossing by foot and bus into neighboring Columbia and other countries in one of the largest migrations in the world today. Many of the migrants live in crowded apartments in Bolivia, which has the bulk of Columbia's virus cases and work as street vendors—jobs now prohibited.

7.5 SIR Model

We present the SIR Model as described in [156] and in [18].

Let $T = [0, \infty)$ and let $\mathcal{P}$ denote the population under study. Let S, I, R be functions of T into the nonnegative integers. Let $t \in T$. We let $S(t)$ denote the number of susceptible persons at time t, $I(t)$ is the number of infected persons at time t, and $R(t)$ is the number of recovered persons at time t. Let $N = |\mathcal{P}|$, i.e., $\mathcal{P}$ has N people.

Let β denote the contact rate and $\frac{1}{\gamma}$ denote the average infectious period.

The model equations are

$$\frac{dS(t)}{dt} = -\frac{\beta}{N} I(t)S(t), \tag{7.5.1}$$

$$\frac{d(I(t))}{dt} = \frac{\beta}{N} I(t)S(t) - \gamma I(t), \tag{7.5.2}$$

$$\frac{dR(t)}{dt} = \gamma I(t). \tag{7.5.3}$$

Let s, i, r be functions of T into $[0, 1]$ defined by for all $t \in T$,

$$s(t) = S(t)/N,$$
$$i(t) = I(t)/N,$$
$$r(t) = R(t)/N.$$

For all $t \in T$, $N = S(t) + I(t) + R(t)$. Thus for all $t \in T$, $1 = s(t) + i(t) + r(t)$ and so $0 = \frac{ds(t)}{dt} + \frac{di(t)}{dt} + \frac{dr(t)}{dt}$.

In [156], b is a fixed number and denotes the number of contacts an infected individual has per day. In [156], k is a fixed fraction of the infected group that will recover during any given day. Thus $b = \beta$ and $k = \gamma$.

Consider Eq. (7.5.2). Factoring out $I(t)$, we have $\frac{d(I(t))}{dt} = (\frac{\beta}{N}S(t) - \gamma)I(t)$. We wish to determine when $\frac{d(I(t))}{dt} = 0$. If $I(t) \neq 0$, then $\frac{d(I(t))}{dt} = 0$ when $\frac{\beta}{N}S(t) - \gamma = 0$ or when $S(t) = \frac{\gamma N}{\beta}$. Solve this equation for t.

Let $k = \frac{1}{3}$ and $b = 4$. Then I has a maximum at some $\widehat{t}$ if $S(\widehat{t}) = \frac{1}{12}N$.

Let $S(0) = S_0, I(0) = I_0$, and $R(0) = R_0$ denote the initial conditions. From Eqs. (7.5.1) and (7.5.3), we have

$$\frac{d(S(t))}{dt} = -\frac{\beta}{N\gamma}\frac{d(R(t))}{dt}S(t) \text{ or}$$
$$\frac{1}{S(t)}\frac{d(S(t))}{dt} = -\frac{\beta}{N\gamma}\frac{d(R(t))}{dt}.$$

By [18], we have

$$S(t) = S_0 e^{-\frac{\beta}{N\gamma}(R(t)-R_0)}. \tag{7.5.4}$$

Since $N = S(t) + I(t) + R(t)$ for all $t \in T$, we have by Eq. (7.5.4) that

$$I(t) = N - S_0 e^{-\frac{\beta}{N\gamma}(R(t)-R_0)} - R(t). \tag{7.5.5}$$

Thus

$$s(t) = \frac{1}{N}S_0 e^{-\frac{\beta}{\gamma}(r(t)-\frac{R_0}{N})}, \tag{7.5.4'}$$

and

$$i(t) = 1 - \frac{1}{N}S_0 e^{-\frac{\beta}{\gamma}(r(t)-\frac{R_0}{N})} - r(t). \tag{7.5.5'}$$

7.6 Intuitionistic Fuzzy Sets

The notion of an intuitionistic fuzzy set was introduced by K. T. Atannasov in [15]. We use it here to introduce a possible area in which to study the coronavirus pandemic.

Definition 7.6.1 Let X be a set. Let $A = \{(x, \mu_A(x), \nu_A(x))\}|x \in X\}$, where μ_A, ν_A are fuzzy subsets of X such that $\mu_A(x) + \nu_A(x) \leq 1$. Then A is called an **intuitionistic fuzzy set (IFS)**.

Let $A = \{(x, \mu_A(x), \nu_A(x))\}|x \in X\}$ be an IFS. Define the fuzzy subset π_A of X by $\forall x \in X, \pi_A(x) = 1 - \mu_A(x) - \nu_A(x)$. In fuzzy set theory, $\mu_A(x)$ is often thought of the degree of membership, $\nu_A(x)$ the degree of nonmembership, and $\pi_A(x)$ the degree of uncertainty of x in a fuzzy subset of X described by A.

We now present a model based on the SIR model as described in the previous section. Let s, i, r be functions from T into $[0, 1]$ such that for all $t \in T$,

$$1 = s(t) + i(t) + r(t).$$

Then $\{(t, s, i, r)|t \in T\}$ is an intuitionistic fuzzy set. We assume that s, i, and r are differentiable. Then

$$\frac{d(i(t))}{dt} = -\frac{ds(t))}{dt} - \frac{d(r(t))}{dt}.$$

We see that $\frac{d(i(t))}{dt} = 0$ if and only if $\frac{ds(t))}{dt} = -\frac{d(r(t))}{dt}$.

Let $\mathcal{R}$ denote the set of all strictly increasing function from T into $[0, 1]$ and $\mathcal{S}$ the set of all strictly decreasing functions from T into $[0, 1]$. Let f be a strictly decreasing function from $[0, 1]$ into itself. For all $r \in \mathcal{R}$. Define the function s from T into $[0, 1]$ by $\forall t \in T, s(t) = f(r(t))$. Then it follows that $s \in \mathcal{S}$. Note that if $t_1, t_2 \in [0, 1]$ with $t_1 < t_2$, then $r(t_1) < r(t_2)$ and so $f(r(t_1)) > f(r(t_2))$. That is, s is strictly decreasing. For a connection with the SIR model, an example of f is e^{-x}, where $x \in [0, 1]$. See (4′).

When experts wish to determine when a virus might peak, they often examine more than one model. In what follows, we provide mathematical methods to combine models in an intuintionistic fuzzy set setting.

Definition 7.6.2 ([15]) Let $A = \{(x, \mu_A(x), \nu_A(x))\}|x \in X\}$ and $B = \{(x, \mu_B(x), \nu_B(x))\}|x \in X\}$ be intuitionistic fuzzy sets.

(1) $A \subseteq B$ if and only if$\forall x \in X, \mu_A(x) \leq \mu_B(x)$ and $\nu_A(x) \geq \nu_B(x)$;
(2) $A \supseteq B$ if and only if $A \subseteq B$;
(3) $A = B$ if and only if $\forall x \in X, \mu_A(x) = \mu_B(x)$ and $\nu_A(x) = \nu_B(x)$;
(4)$\overline{A} = \{(x, \nu_A(x), \mu_A(x))\}|x \in X\}$.

Definition 7.6.3 ([15]) Let $A = \{(x, \mu_A(x), \nu_A(x))\}|x \in X\}$ and $B = \{(x, \mu_B(x), \nu_B(x))\}|x \in X\}$ be intuitionistic fuzzy sets.

(1) $A \cap B = \{(x, \mu_A(x) \wedge \mu_B(x), \nu_A(x) \vee \nu_B(x))|x \in X\}$;
(2) $A \cup B = \{(x, \mu_A(x) \vee \mu_B(x), \nu_A(x) \wedge \nu_B(x))|x \in X\}$;
(3) $A + B = \{(x, \mu_A(x) + \mu_B(x) - \mu_A(x)\mu_B(x), \nu_A(x)\nu_B(x))|x \in X\}$;
(4) $A \cdot B = \{(x, \mu_A(x)\mu_B(x), \nu_A(x) + \nu_B(x) - \nu_A(x)\nu_B(x))|x \in X\}$;
(5) $A@B = \{(x, \frac{\mu_A(x)+\mu_B(x)}{2}, \frac{\nu_A(x)+\nu_B(x)}{2})|x \in X\}$;
(6) $A\$B = \{(x, \sqrt{\mu_A(x)\mu_B(x)}, \sqrt{\nu_A(x)\nu_B(x)})|x \in X\}$;
(7) $A * B = \{(x, \frac{\mu_A(x)+\mu_B(x)}{2(\mu_A(x)\mu_B(x)+1)}, \frac{\nu_A(x)+\nu_B(x)}{2(\nu_A(x)\nu_B(x)+1)})|x \in X\}$;

$$(8) A \circledast B = \begin{cases} \{(x, 2\frac{\mu_A(x)\mu_B(x)}{\mu_A(x)+\mu_B(x)}, 2\frac{\nu_A(x)\nu_B(x)}{\nu_A(x)+\nu_B(x)})|x \in X\} \text{ if } \mu_A(x) + \mu_B(x) \neq 0 \neq \nu_A(x) + \nu_B(x), \\ \{(x, 0, 2\frac{\nu_A(x)\nu_B(x)}{\nu_A(x)+\nu_B(x)})|x \in X\} \text{ if } \mu_A(x) + \mu_B(x) = 0 \neq \nu_A(x) + \nu_B(x), \\ \{(x, 2\frac{\mu_A(x)\mu_B(x)}{\mu_A(x)+\mu_B(x)}, 0)|x \in X\} \text{ if } \mu_A(x) + \mu_B(x) \neq 0 = \nu_A(x) + \nu_B(x), \\ \{(x, 0, 0)|x \in X\} \text{ if } \mu_A(x) + \mu_B(x) = 0 = \nu_A(x) + \nu_B(x). \end{cases}$$

Theorem 7.6.4 ([15]) *The sets given in (1)–(8) of Definition 7.6.3 are intuitionistic fuzzy sets.*

We are interested in intuitionistic fuzzy sets $\{(t, s(t), r(t))|t \in T\}$, where s is strictly decreasing and r is strictly increasing. Let $\mathcal{I}$ denote the set of all such sets.

Proposition 7.6.5 *Suppose* $A = \{(t, s_A(t), r_A(t))|t \in T\}$, $B = \{(t, s_B(t), r_B(t))|t \in T\} \in \mathcal{I}$. *Then*
(1) $A \cap B \in \mathcal{I}$;
(2) $A \cup B \in \mathcal{I}$;
(3) $A + B \in \mathcal{I}$;
(4) $A \cdot B \in \mathcal{I}$;
(5) $A@B \in \mathcal{I}$;
(6) $A\$B \in \mathcal{I}$.

Proof Let $t_1, t_2 \in T$ be such that $t_1 < t_2$.

(1) Then $(s_A \wedge s_B)(t_1) = s_A(t_1) \wedge s_B(t_1) > s_A(t_2) \wedge s_B(t_2) = (s_A \wedge s_B)(t_2)$. Hence $s_A \wedge s_B$ is strictly decreasing. Also, $(r_A \vee r_B)(t_1) = r_A(t_1) \vee r_B(t_1) < r_A(t_2) \vee r_B(t_2) = (r_A \vee r_B)(t_2)$. Thus $r_A \vee r_B$ is strictly increasing.

(2) We have that $(s_A \vee s_B)(t_1)=s_A(t_1) \vee s_B(t_1) > s_A(t_2) \vee s_B(t_2)=(s_A \vee s_B)(t_2)$. Hence $s_A \vee s_B$ is strictly decreasing. Also, $(r_A \wedge r_B)(t_1)=r_A(t_1) \wedge r_B(t_1) < r_A(t_2) \wedge r_B(t_2)=(r_A \wedge r_B)(t_2)$. Thus $r_A \wedge r_B$ is strictly increasing.

(3) We have that $s_A(t_1) > s_A(t_2)$ and $s_B(t_1) > s_B(t_2)$. Thus $s_A(t_2) \neq 1 \neq s_B(t_2)$. It is suffices to show for $a, b, c \in [0, 1]$ with $a < c$ and $b \neq 1$ that $a + b - ab < c + b - cb$. This follows since $a(1 - b) < c(1 - b)$. We also have $r_A(t_1) < r_A(t_2)$ and $r_B(t_1) < r_B(t_2)$. Clearly, $r_A(t_1)r_B(t_1) < r_A(t_2)r_B(t_2)$.

(4) The proof is similar to that of (3).

(5) and (6) Immediate. □

We note that if s_A, r_A, s_B, r_B are differentiable, then so are the successor and recovery functions in (3)–(6).

We also note that in (1)–(4) involve t-norms and t-conorms, while (5) involves an aggregation operator.

Proposition 7.6.6 *Let* $A= \{(t, s_A(t), r_A(t))|t \in T\}$, $B=\{(t, s_B(t), r_B(t))|t \in T\} \in \mathcal{I}$. *Suppose that* $f : [0, 1] \to [0, 1]$ *is strictly decreasing and* $r_A, r_B \in \mathcal{R}$. *Let* $s_A(t) = f(r_A(t))$ *and* $s_B(t) = f(r_B(t))$ *for all* $t \in T$. *Let* $t \in T$. *Suppose* $r_A(t) \leq r_B(t)$ *implies* $s_A(t) \geq s_B(t)$. *Then*
(1) $f(r_{A\cup B})(t) = s_{A\cap B}(t)$,
(2) $f(r_{A\cap B})(t) = s_{A\cup B}(t)$.

Proof (1) Recall $A \cap B = \{(t, s_{A\cap B}, r_{A\cup B})|t \in T\}$. Suppose $r_A(t) \leq r_B(t)$. Then $f(r_{A\cup B}(t)) = f((r_A(t) \vee r_B(t)) = f(r_B(t))$ and $s_{A\cap B}(t) = s_A(t) \wedge s_B(t) = s_B(t) = f(r_B(t))$. Thus $f(r_{A\cup B})(t) = s_{A\cap B}(t)$.

(2) Recall $A \cup B = \{(t, s_{A\cup B}, r_{A\cap B})|t \in T\}$. Suppose $r_A(t) \leq r_B(t)$. then $f(r_{A\cap B})(t)) = f((r_A(t) \wedge r_B(t))=f(r_A(t))$ and $s_{A\cup B}(t) = s_A(t) \vee s_B(t) = s_A(t) = f(r_A(t))$. Thus $f(r_{A\cap B})(t) = s_{A\cup B}(t)$. □

7.7 SEIR Model

This model is based on a Susceptible-Exposed-Infectious-Recovered (SEIR) model to capture the natural history of COVID-19 and its transmission dynamics, [86]. See [67] also.The following are the governing equations:

$$\frac{dS}{dt} = -\lambda S,$$
$$\frac{dE}{dt} = \lambda S - rE,$$
$$\frac{dI}{dt} = R - \gamma I - \mu I,$$
$$\lambda = \frac{\beta I}{N} + \frac{k\beta E}{n},$$

where $S(t)$ is the number of susceptible persons at time t, $E(t)$ is the number of persons exposed and infectious, but not yet symptomatic at time t, $I(t)$ is the number of infected and symptomatic persons at time t, and $R(t)$ is the number of recovered persons at time t. Then S, E, I, R are functions from T into the set of nonnegative integers. Here N denotes the size of the population. The model parameters are exposed, per-capita rate of developing symptoms r; among symptomatic, per-capita rates of recovery and death (γ and μ, respectively) and the average number of infections caused per day per symptomatic case β and the infectiousness of exposed/asymptomatic cases, relative to symptomatic k.

The population in each metropolitan area is divided into different compartments, representing states of disease, with flows into different compartments, representing states of disease, with flows between compartments given by certain rates. Thus susceptible individuals upon acquiring infection enter a state of asymptomatic infection and with some delay develop symptomatic disease. It is assumed that a proportion p of symptomatic cases is subject to quarantine $I^{(q)}$ and the remainder $I^{(n)}$ is not. The relative size of these two populations p reflects the coverage of quarantine delay $\frac{1}{\delta}$. Individuals may be cured or die as per recovery rate γ or mortality rate μ, respectively. Those people who are successfully quarantined Q do not contribute to onward infection.

Chapter 8
Fuzzy Graphs and Human Trafficking

Bottleneck problems, largely studied in optimization and quality of service problems, depends directly on the connectivity between pairs of nodes. Connectivity is one of the major themes of fuzzy graph theory. Both the introductory papers written in 1975 by Rosenfeld [143] and Yeh and Bang [171] were focussed at this theme. Later fuzzy vertex connectivity κ, fuzzy edge connectivity κ' and t-connected fuzzy graphs were introduced in [98]. This chapter primarily focusses on containers and spanning containers, constructions of t-connected fuzzy graphs, average fuzzy vertex connectivity and uniformly connected fuzzy graphs.

In the following section, *containers* in fuzzy graphs are discussed. A container in a graph is a family of internally disjoint paths between a pair of vertices. Analogously, a container in a fuzzy graph is a collection of internally disjoint strongest paths. Strongest paths always contribute the maximum towards the flow in a network. Hence the study of strongest paths and containers helps in analyzing and controlling the dynamics of networks. In this section, two classic results on graph degrees are generalized to fuzzy graphs (Theorems 8.1.19 and 8.1.21). Spanning containers of blocks, fuzzy trees and locamin cycles are discussed. An algorithm to identify maximum and minimum container capacities and an application related to human trafficking are proposed.

8.1 Containers in Fuzzy Graphs

As mentioned before, more number of widest paths between a particular pair of locations in a network can provide multiple facilities between those locations. Hence the study of collection of paths between pairs of nodes becomes very relevant. In the following section, we look at a collection of strongest paths between pairs of vertices instead of single strongest paths between them. We concentrate on undirected fuzzy

J. N. Mordeson and S. Mathew, *Mathematics of Uncertainty for Coping with World Challenges*, Studies in Systems, Decision and Control 353,
https://doi.org/10.1007/978-3-030-68684-0_8

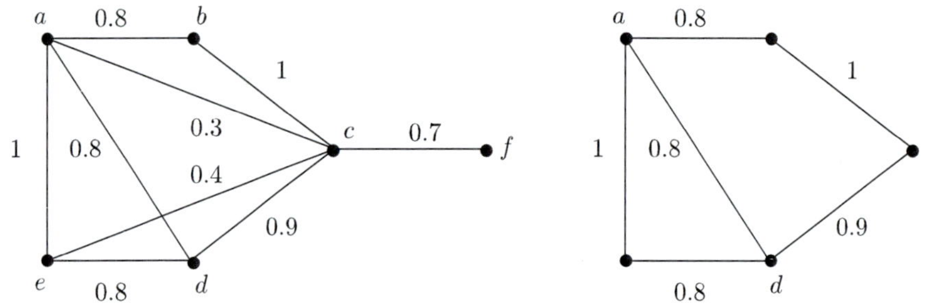

Fig. 8.1 A fuzzy graph and its unique (a, d) 3-fuzzy container

graphs and vertex disjoint strongest paths. The definition of a k-fuzzy container between a pair of vertices is given below.

Definition 8.1.1 Let u, v be any two vertices of a fuzzy graph $G = (\sigma, \mu)$. A set of k internally disjoint strongest paths between u and v is said to be a *k-fuzzy container* of G between u and v, denoted by $C_G^k(u, v)$. Also, denote $C_G^k(u, v)$ as $C_G(u, v)$ if k is the maximum number of internally disjoint strongest $u - v$ paths in G.

Thus, $C_G^k(u, v)$ consists of paths $P_1 = (\tau_1, \nu_1)$, $P_2 = (\tau_2, \nu_2), \ldots, P_k = (\tau_k, \nu_k)$ such that $\nu_i^* \cap \nu_j^* = \emptyset$ and $\tau_i^* \cap \tau_j^* = \{u, v\}$ for $1 \leq i \neq j \leq k$.

For convenience, we say that there exists a 0-container between u and v if u and v are not connected by a path in G. If G contains a k-fuzzy container between u and v, then it contains an l-fuzzy container between u and v for $1 \leq l \leq k$. If uv is an α-strong edge, then uv is a unique 1-fuzzy container between u and v in G since no other $u - v$ strongest paths exists. We can obtain the definition of a less complicated family of paths in G, named as a strong container if we replace strongest paths by strong paths in Definition 8.1.1.

Example 8.1.2 Let the fuzzy graph $G = (\sigma, \mu)$ has vertex set $\sigma^* = \{a, b, c, d, e, f\}$, with edge membership values $\mu(ab) = \mu(ad) = \mu(ed) = 0.8$, $\mu(bc) = \mu(ae) = 1$, $\mu(cd) = 0.9$, $\mu(ce) = 0.4$, $\mu(ac) = 0.3$ and $\mu(cf) = 0.7$ (See Fig. 8.1). In this graph, $P_1 = a - b - c - d$; $P_2 = a - d$ and $P_3 = a - e - d$ are three strongest internally disjoint $a - d$ paths. There are many fuzzy containers in G. $C_G^3(a, d) = \{P_1, P_2, P_3\}$ is a 3-fuzzy container. Also $C_G^2(a, d) = \{P_1, P_2\}$ and $C_G^1(a, d) = \{P_1\}$ are 2-fuzzy containers. Note that there does not exist an (a, d) 4-fuzzy container in G. All containers in G are strong. Consider the vertices $b, d \in \sigma^*$. There are two $b - d$ strong paths (namely $b - a - e - d$ and $b - c - d$) and a unique $b - d$ strongest path (that is $b - c - d$) in G. Hence, there exists a strong 2- container and a 1-fuzzy container between b and d in G.

The next proposition is about the existence of containers and strong containers in fuzzy graphs. Even though the proof is straight forward, we present it for a better understanding of the concepts.

Proposition 8.1.3 *There always exists at least one fuzzy container and at least one strong container between every pair of vertices in a connected fuzzy graph.*

Proof Let $G = (\sigma, \mu)$ be a connected fuzzy graph. Let T be a maximum spanning tree of G. By Theorem 1 of [150], every edge in T is strong. Also every $u - v$ path in T is a strongest $u - v$ path in G. Thus there exists a 1-fuzzy container as well as a strong container between every pair of vertices. ■

Next we have an easy characterization for fuzzy cycles using containers.

Theorem 8.1.4 *Let $G = (\sigma, \mu)$ be a fuzzy graph such that its underlying graph G^* is a cycle. Then G is a fuzzy cycle if and only if there exists a 2-fuzzy container in G.*

Proof Assume that G is a fuzzy cycle. Let uv be a weakest edge of G. Let P_1 be the edge uv and $P_2 = G - uv$. Then $s(P_1) = s(P_2) = CONN_G(u, v)$. Hence, $C_G(u, v) = \{P_1, P_2\}$ is a 2-fuzzy container in G.

Conversely, assume that the condition of the theorem is true. If G is not a fuzzy cycle, then there exists exactly one weakest edge in G. Since G^* is a cycle, it follows that all containers in G are 1-fuzzy containers in G, which is a contradiction. ■

Corollary 8.1.5 *Let $G = (\sigma, \mu)$ be a fuzzy graph with $|\sigma^*| = n$. Then G is a fuzzy cycle if and only if there exist $\binom{n}{2} - 2p$ pairs of 2-fuzzy containers in G, where p is the number of α-strong edges of G.*

In any fuzzy cycle $G = (\sigma, \mu)$, for any two vertices $u, v \in \sigma^*$, there always exist at most two distinct 1-fuzzy containers and a unique strong 2-container between u and v. A fuzzy tree may contain more than one distinct strongest paths. Also, every strong path in a fuzzy tree is a strongest path, while the converse is not true. The following theorem is a characterization of fuzzy forests using the concept of k-fuzzy containers.

Theorem 8.1.6 *$G = (\sigma, \mu)$ is a fuzzy forest if and only if there exist at most one 1-fuzzy container between every pair of vertices of G.*

Proof Assume that G is a fuzzy forest. Then there exist at most one internally disjoint strongest path between any two vertices of G. Let u, v belong to the same component G' of G and F' be the unique maximal spanning tree of G'. Every path in F' is a strongest path, which forms a strongest path of G' and hence of G. The vertices u, v lies in F' and thus there exists a $u - v$ strongest path P' in G. Hence there exists a unique 1-fuzzy container between u and v. If u and v belong to different components of G, then there does not exist a $u - v$ path in G. Hence we say that there exists a 0-container between u and v in G.

We prove the converse by contradiction. Assume that G is not a fuzzy forest. Then there exists a component G' with a fuzzy cycle say C. By Theorem 8.1.4, there exists a 2-fuzzy container between some pair of vertices in C. That is, there exist two distinct 1-fuzzy containers between those pairs of vertices in G' and hence in G, which violates the statement of the theorem. ∎

Corollary 8.1.7 *Let G be a fuzzy graph. Then G is a fuzzy tree if and only if there exists a unique fuzzy container between every pair of vertices in G.*

Extraction of a k-fuzzy containers from a complete fuzzy graph is complicated. But we have a theorem for finding the number of 1-fuzzy containers in a complete fuzzy graph.

Theorem 8.1.8 *Let* $G = (\sigma, \mu)$ *be a complete fuzzy graph with* $|\sigma^*| = n > 2$. *If* $\sigma(v_1) < \sigma(v_2) < \cdots < \sigma(v_{n-1}) < \sigma(v_n)$, *then there exist* $m = \sum_{p=1}^{n-1}(n-p)\left[\sum_{i=0}^{n-p-1}\binom{n-p-1}{i}\right]$ *distinct* 1*-fuzzy containers in G.*

Proof Since $\sigma(v_1) < \sigma(v_2) < \cdots < \sigma(v_{n-1}) < \sigma(v_n)$, every edge $uv \in \mu^*$ is a strongest $u - v$ path in G. Between v_1 and v_j, there exist exactly $\binom{n-2}{i}$ strongest paths with i internal vertices. Hence for the $n-1$ pairs of vertices $\{v_1, v_j\}_{j=2}^{n}$, there exists

$$\sum_{i=0}^{n-2}\binom{n-2}{i}$$

1-fuzzy containers in G. And for the $n-2$ pairs of vertices $\{v_2, v_j\}_{j=3}^{n}$, there exists

$$\sum_{i=0}^{n-3}\binom{n-3}{i}$$

1-fuzzy containers in G. In general, for the $n-k$ pairs of vertices $\{v_k, v_j\}_{j=k+1}^{n}$, there exists

$$\sum_{i=0}^{n-k-1}\binom{n-k-1}{i}$$

1-fuzzy containers in G. Summing up all these numbers we get,

$$m = \sum_{p=1}^{n-1}(n-p)\left[\sum_{i=0}^{n-p-1}\binom{n-p-1}{i}\right]$$

distinct 1-fuzzy containers in G. ∎

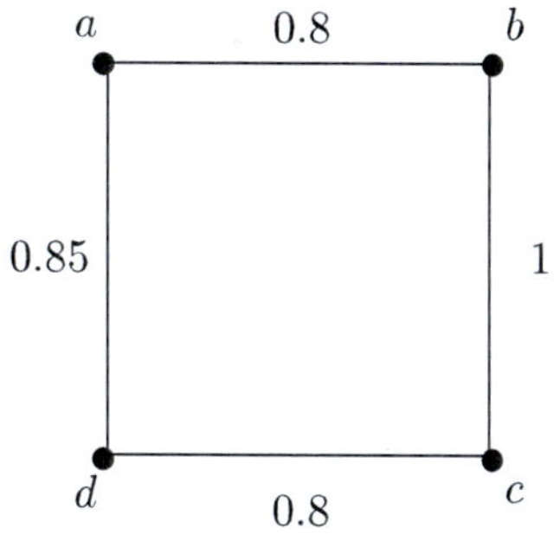

Fig. 8.2 A 2^*-connected fuzzy cycle

Definition 8.1.9 A fuzzy graph $G = (\sigma, \mu)$ is said to be k^**-connected* if there exist t containers $C_G^t(u, v)$, $1 \leq t \leq k$ for every pair of vertices $u, v \in \sigma^*$ such that uv is not an α-strong edge.

Example 8.1.10 Let $G = (\sigma, \mu)$ be the fuzzy graph with $\sigma^* = \{a, b, c, d\}$, $\mu(ab) = \mu(cd) = 0.8$, $\mu(bc) = 1$ and $\mu(ad) = 0.85$ (Fig. 8.2). Then, $C_G^2(a, b) = \{a - b, a - d - c - b\}$, $C_G^2(c, d) = \{c - d, c - b - a - d\}$, $C_G^2(a, c) = \{a - b - c, a - d - c\}$ and $C_G^2(b, d) = \{b - a - d, b - c - d\}$. Thus G is 2^*-connected.

Note that a fuzzy cycle is always 1^*-connected and can be even 2^*-connected, while a fuzzy tree with a cyclic support is 1^*-connected, but not 2^*-connected. Clearly, any k^*-connected fuzzy graph, where $k \geq 2$ is a block. Also there exist k internally disjoint strong paths between any two vertices u, v when the fuzzy graph is k^*-connected. In short, a strong container can be always extracted from a fuzzy container as seen from the following result.

Theorem 8.1.11 *If G is a k^*-connected fuzzy graph, then $|N_s(u)| \geq k$ for any $u \in \sigma^*$.*

Proof Let $G = (\sigma, \mu)$ be a k^*-connected fuzzy graph. The theorem is obvious when $k = 1$. Let $u \in \sigma^*$. Now choose any arbitrary vertex $v \in \sigma^*$. If edge uv is a not a fuzzy bridge, then by assumption there exist $k \geq 2$ internally disjoint strongest $u - v$ paths in G. It is clear that there are k-strong paths starting at u. Thus $|N_s(u)| \geq k$. Now suppose that uv is a fuzzy bridge. Then uv is the only strongest $u - v$ path in G. Let $w \in N(u)$. Clearly the edge uw is not a fuzzy bridge; for otherwise, by Theorem 2.2.10 of [90] u becomes a fuzzy cutvertex, which is a contradiction to the fact that G is 2^*-connected and hence is a block. Hence by the first part of the proof it follows that $|N_s(u)| \geq k$. ∎

The converse of the above theorem need not be true. Consider the fuzzy graph G given in Fig. 8.4. Here, $|N_s(y)| \geq 2$ for every $y \in \sigma^*$. But, x is a cutvertex and hence G is not 2^* connected.

If G is a block and $|N_s(v)| \geq k$ for each $v \in \sigma^*$, then G is k^*-connected. Consequently, strongest strong cycles and locamin cycles are 2^*-connected fuzzy graphs.

Proposition 8.1.12 *A connected fuzzy graph $G = (\sigma, \mu)$ where $|\sigma^*| \geq 3$, and having no fuzzy bridges has a 2^*-connected fuzzy subgraph.*

Proof Clearly, all edges in μ^* are either δ-edges or β-strong edges. Choose a β-strong edge $uv \in \mu^*$. Then there exists a strongest $u - v$ path P different from the edge uv, whose strength is $\mu(uv)$. Let $H = P \cup uv$, which is a strongest strong cycle. H is a 2^*-connected fuzzy subgraph of G. ■

Theorem 8.1.13 *Let $G = (\sigma, \mu)$ be a 2^*-connected fuzzy graph. Then there exists a sequence $\{B_i = (\sigma_i, \mu_i)\}_{i=1}^{l}$ of 2^*-connected fuzzy subgraphs with $\sigma_l^* = \sigma^*$ and edge disjoint paths $\{P_i = (\rho_i, \nu_i)\}_{i=1}^{l-1}$ such that B_{i+1} is a fuzzy subgraph induced by $\sigma_i \cup \rho_i$ and $|\sigma_i \cap \rho_i| = 2$ (only the end vertices of P_i) for $0 < i \leq l - 1$.*

Proof Suppose that G is a 2^*-connected fuzzy graph. We arbitrarily choose a 2^*-connected fuzzy subgraph $B_i = (\sigma_i, \mu_i)$ in G (see Fig. 8.3). If $\sigma_i^* = \sigma^*$, then we are done. Suppose $\sigma_i^* \neq \sigma^*$, then there exists a strong edge $uv \in \mu^*$ such that the vertex $u \in \sigma_i^*$ but the edge $uv \notin \mu_i^*$. If v also belongs to σ_i^*, take $P_i = uv$ and set $\sigma_{i+1} = \sigma_i \cup \{u, v\}$. Let B_{i+1} be the fuzzy subgraph induced by σ_{i+1}^*. Note that uv is a β-strong edge (since $u, v \in \sigma_i^*$). Clearly B_{i+1} is 2^*-connected.

Suppose uv is a strong edge of G having only one of its ends, say u, in σ_i^*. Let w be a vertex other than u in σ_i^*. Then by Theorem 2 of [94], since G is 2^*-connected, there is a cycle C_i containing the vertex w and the strong edge uv formed by two strongest strong $w - u$ or $w - v$ paths. Let w_i be the first vertex of B_i in the $v - w$ section of C_i containing v, and let $P_i = (\rho_i, \nu_i)$ be the respective $u - w_i$ section of C_i. Set $\sigma_{i+1} = \sigma_i \cup \rho$. Letting B_{i+1} as the fuzzy subgraph induced by σ_{i+1}. By construction B_{i+1} is 2^*-connected.

Applying induction on i, we obtain a sequence of $\{B_i = (\sigma_i, \mu_i)\}_{i=1}^{l}$ of 2^*-connected fuzzy subgraphs with $\sigma_l^* = \sigma^*$ and edge disjoint paths $\{P_i = (\rho_i, \nu_i)\}_{i=1}^{l-1}$ such that B_{i+1} is a fuzzy subgraph induced by $\sigma_i \cup \rho_i$ for $0 < i \leq l - 1$. ■

Definition 8.1.14 Let $G = (\sigma, \mu)$ be a fuzzy graph. Then the capacity of a fuzzy container $C_G^k(u, v)$ denoted by $Cap(C_G^k(u, v))$ is defined as $k \times CONN_G(u, v)$.

Definition 8.1.15 The minimum and maximum container capacity of a fuzzy graph G is defined as

$$minCap(G) = \min\{\, Cap(C_G(u, v)) : u, v \in \sigma^*, uv \text{ is not a fuzzy bridge } \}$$
$$maxCap(G) = \max\{\, Cap(C_G(u, v)) : u, v \in \sigma^*, uv \text{ is not a fuzzy bridge } \}$$

Consider a fixed pair of vertices u, v. The maximum capacity of a container between u and v in G would be $k \times CONN_G(u, v)$, where k is the maximum number of internally disjoint strongest $u - v$ paths in G.

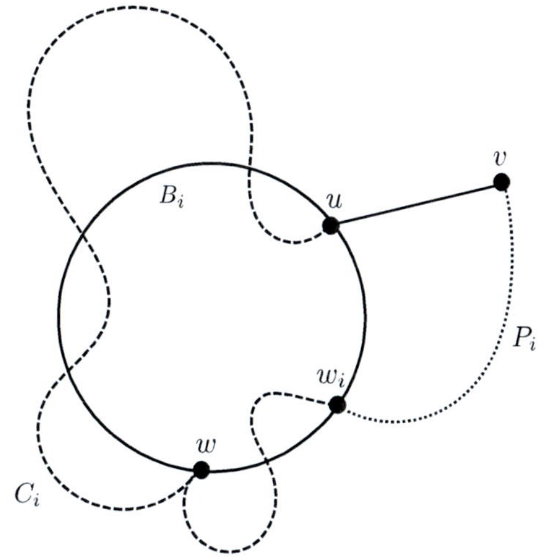

Fig. 8.3 Construction of B_{i+1} in the proof of Theorem 8.1.13

If $G = (\sigma, \mu)$ is a fuzzy tree, then, $Cap(C^k_G(u, v)) \geq \kappa(G)$ for every $u, v \in \sigma^*$. Also, a container with minimum capacity, $C^k_G(x, y)$ in a fuzzy tree is the strong edge $xy \in \mu^*$ with minimum membership value. Hence $minCap(G) = Cap(C_G(x, y)) = \kappa(G)$. We have an obvious result.

Theorem 8.1.16 *The fuzzy vertex connectivity $\kappa(G)$ of a fuzzy graph G is the minimum container capacity of G.*

Theorem 8.1.17 *Let $G = (\sigma, \mu)$ be a complete fuzzy graph with $|\sigma^*| = n \geq 3$. If $\sigma(v_1) \leq \sigma(v_2) \leq \cdots \leq \sigma(v_{n-1}) \leq \sigma(v_n)$, then $minCap(G) = \kappa(G)$ if $\dfrac{\sigma(v_1)}{\sigma(v_i)} \leq \dfrac{n-i}{n-1}$ for $i = 2, 3, \ldots, n-2$, else $minCap(G) < \kappa(G)$.*

Proof Let G be a complete fuzzy graph such that $\frac{\sigma(v_1)}{\sigma(v_i)} \leq \frac{n-i}{n-1}$ for $i = 2, 3, \ldots, n-2$. Then,

$$(n-1)\sigma(v_1) \leq (n-i)\sigma(v_i) \text{ for } i = 2, 3, \ldots, n-2 \ldots. \tag{8.1.1}$$

Using (1), we have,

$$\begin{aligned} Cap(C_G(v_1, v_i)) &= (n-1)\sigma(v_1) \text{ for } i = 2, 3, \ldots, n. \\ &\leq (n-i)\sigma(v_i) \text{ for } i = 2, 3, \ldots, n-2 \ldots. \end{aligned} \tag{8.1.2}$$

for $i < j$, $i \in \{2, 3, \ldots, n-2\}$ and $j \in \{3, 4, \ldots, n-2\}$,

$$Cap(C_G(v_i, v_j)) = Cap(C_G(v_i, v_{i+1})) \geq (n-i)\sigma(v_i), \ldots \tag{8.1.3}$$

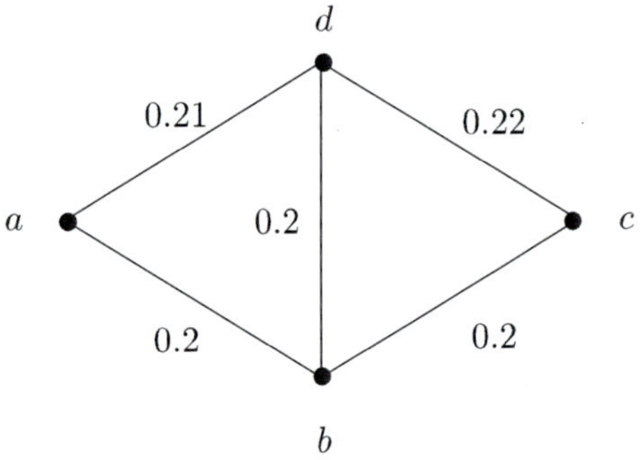

Fig. 8.4 A fuzzy graph G with $\kappa(G) < minCap(G)$

From Eqs. (8.1.2) and (8.1.3), for any pair of vertices v_j, v_l, $Cap(C_G(v_j, v_k)) \geq Cap(C_G(v_1, v_i))$ for $i = 2, 3, \ldots, n-2$. That is, $C_G(v_1, v_i)$ is a fuzzy container which gives minimum container capacity of G. Moreover, $\kappa(G) = (n-1)\sigma(v_1)$. Hence, $minCap(G) = \kappa(G)$. ■

Theorem 8.1.18 *Let G=(σ, μ) be a connected fuzzy graph. Then* $\wedge\{Cap(C_G(u, v)) : u, v \in \sigma^* and uv \text{is not strong}\} \geq \kappa(G)$.

Proof Let G be a connected fuzzy graph. Choose a pair of vertices u, v such that uv is not a strong. Let $C_G(u, v) = \{P_1, P_2, \ldots, P_k\}$ for some $k \geq 1$. It is clear that $s(P_i) = s(P_j)$, $i \neq j$ and there exists an edge $u_i v_i$ in P_i such that $s(P_i) = \mu(u_i v_i)$. Set

$$x_i = \begin{cases} u_i, & \text{if } u_i \neq u \\ v_i, & \text{if } v_i \neq v \end{cases}.$$

Let $A = \{x_1, x_2, \ldots, x_k\}$. Since A forms a fuzzy vertex cut, $CONN_{G-A}(u, v) < CONN_G(u, v)$. Hence $s(A) \geq \kappa(G)$. Here the strong weight of A, $s(A)$ is actually $Cap(C_G(u, v))$. Thus, $Cap(C_G(u, v)) \geq \kappa(G)$. Since u, v are arbitrary, it follows that $\wedge\{Cap(C_G(u, v)) : u, v \in \sigma^* and uv \text{is not strong}\} \geq \kappa(G)$. ■

The following example shows that fuzzy vertex connectivity $\kappa(G)$ of a fuzzy graph strictly exceeds the minimum capacity of a fuzzy graph.

Consider the fuzzy graph $G = (\sigma, \mu)$ with $\sigma^* = \{a, b, c, d\}$, $\mu(ab) = \mu(bc) = \mu(bd) = 0.2$, $\mu(ad) = 0.2$ and $\mu(cd) = 0.22$ (Fig. 8.4). Here, $\kappa(G) = 0.2$ and minimum container capacity of G is $Cap(C_G(a, c)) = 0.21$, which is strictly greater than $\kappa(G)$.

Theorem 8.1.19 *Let $G = (\sigma, \mu)$ be a fuzzy graph without fuzzy cutvertices. If G contains a (u, v) 1-fuzzy container $C_G(u, v)$ with $S(C_G(u, v)) > t$, where uv is not an α-strong edge, then G contains a fuzzy cycle of size greater than $2t + \epsilon$, $\epsilon > 0$.*

Proof Let $P = u_1, u_2, \ldots, u_m$ be the unique strongest $u - v$ path in $C_G(u, v)$. We have $Cap(C_G(u, v)) = CONN_G(u, v) > t$ and uv is not an α-strong edge. Suppose that uv is not strong. Since G contains no fuzzy cutvertices, there exist at least two vertices in a $u - v$ strength reducing set. By Corollary 8.1.7, there exists two strongest $u - v$ paths in G. We denote these paths by P_1 and P_2. The strength of each path is $CONN_G(u, v)$ and size of path P_i always greater than or equal to $CONN_G(u, v)$. Let $S(P_i) = CONN_G(u, v) + \epsilon_i$, $\epsilon_i > 0$, for $i = 1, 2$. Now let $C = P_1 \cup P_2$. Then C is a fuzzy cycle with strength $CONN_G(u, v)$. The size of C is, $S(C) = CONN_G(u, v) + \epsilon_1 + CONN_G(u, v) + \epsilon_2 > t + \epsilon_1 + t + \epsilon_2 = 2t + \epsilon$, where $\epsilon = \epsilon_1 + \epsilon_2$.

Suppose uv is a β-strong edge in G. Then letting P_1 as the edge uv and P_2 as the strongest $u - v$ path other than uv and by taking $C = P_1 \cup P_2$, we can see that $S(C) = CONN_G(u, v) + \epsilon + CONN_G(u, v) > t + \epsilon + t = 2t + \epsilon$. ■

Next we have a fuzzy graph generalization of a result by Dirac [45], who has obtained relationship between the degrees of vertices and lengths of paths in a graph. The relationship between strong degrees of vertices of a fuzzy graph and the sizes of strong paths is provided in Theorem 8.1.20.

Theorem 8.1.20 *Let $G = (\sigma, \mu)$ be a connected fuzzy graph with $|\sigma^*| \geq 2$. Suppose $d_s(v) \geq t$ for every $v \in \sigma^* \setminus \{x\}$ and $\mu(e) > 0$ for every $e \in \mu^*$. Then G contains an $x - y$ strong path with size at least t for some $y \in \sigma^*$. Moreover, if G contains no $x - y$ path of size more than t, then each block $B = (\tau, \nu)$ of G is complete, $x \in \tau^*$ and $\nu(xv) = h_0$ for all $v \in \tau^* \setminus \{x\}$ and $\nu(uv) = h_1$ for all $u, v \in \tau^* \setminus \{x\}$, where h_0 and h_1 are constants.*

Proof We will prove the result by mathematical induction on $|\sigma^*|$. If $|\sigma^*| = 2$, then the result is trivially true. Suppose that $|\sigma^*| > 2$. Let $H = (\sigma', \mu')$, where $\sigma'^* = \sigma^* \setminus \{x\}$ and $\mu'(u, v) = \mu(uv)$ for $u, v \in \sigma^*$.

Suppose H is connected. Let $\epsilon = \max\{\mu(xu) : u \in N_s(x)\}$. Choose a vertex $w \in N_s(x)$ such that $\mu(xw) = \epsilon$. Let $t' = t - \epsilon$. Then $d_s(v) \geq t'$ for all $v \in \sigma^* \setminus x$. By induction hypothesis, H contains a $w - y$ strong path Q of size at least t'. Thus $xw \cup Q$ is a $x - y$ strong path say P in G and size of P is at least $t' + \epsilon = t$.

Suppose H is disconnected. Let $H_1 = (\sigma_1, \mu_1)$, $H_2 = (\sigma_2, \mu_2), \ldots, H_l = (\sigma_l, \mu_l)$ be the components of H. Then apply the induction hypothesis to the fuzzy subgraphs of G induced by $\sigma_i \cup \{x\}$ for each i, $1 \leq i \leq l$ to get the result.

If G contains no $x - y$ strong paths of size more than t (for example refer fuzzy graph shown in Fig. 8.5), then let H be the fuzzy subgraph induced by $\sigma^* \setminus \{x\}$ and $\epsilon = \max\{\mu(xu) : u \in N_s(x)\}$. Choose a vertex $w \in N_s(x)$ such that $\mu(xw) = \epsilon$. Let $t' = t - \epsilon$. Then the size of every $w - y$ strong path in H is less than t' (If H has a $w - y$ strong path in H with size at least t', then extending the path with the edge xw gives an $x - y$ strong path in G with size at least t). Hence by induction hypothesis, each block $B' = (\tau', \nu')$ of H is complete, $w \in \tau'^*$ and $\nu(wv) = h'_0$ for

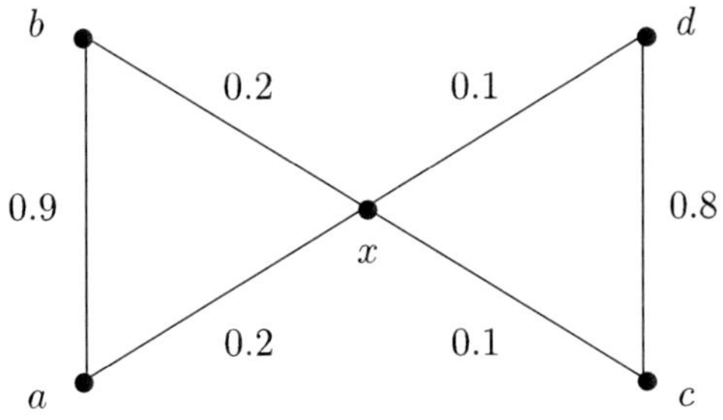

Fig. 8.5 A fuzzy graph with no $x - v$ strong paths whose size is greater than $t = 0.7$

all $v \in \tau'^{*} \backslash \{w\}$ and $\nu(uv) = h'_1$ for all $u, v \in \tau'^{*} \backslash \{w\}$ where h'_0 and h'_1 are constants.

Since $\epsilon > 0, xw \in \mu^*$. Moreover, by choice of w, $\mu(xw) = h_0$. Thus, $\mu(xv) = h_0$ for all $v \in \tau'^*$. It follows that any vertex of B' could have been selected as the vertex w. This immediately implies that $H = B'$ and that $h'_0 = h'_1$. Now set $h_1 = h'_0$, to complete the proof. ■

In the following result, we provide the relationship between the degrees of the vertices of a fuzzy graph and the size of a fuzzy cycle in G.

Theorem 8.1.21 *Let $G = (\sigma, \mu)$ be a connected fuzzy graph with $|\sigma^*| \geq 2$. Suppose $d_s(v) \geq t$ for every $v \in \sigma^* \backslash \{x\}$. If $|N_s(v)| \geq 2$ for every $v \in \sigma^* \backslash \{x\}$, then G contains a fuzzy cycle of size more than t.*

Proof Let $P : x = x_0, x_1, \ldots, x_m$, be a strong path in G of maximum size, and as long as possible subject to these conditions. Then x_m is adjacent only to vertices of P. Let

$$X = \{i : x_i x_m \in \mu^*\} \text{ and } j = \min X.$$

Then for each $i \in X$, $P' : x = x_0, x_1, \ldots, x_i, x_m, x_{m-1}, \ldots, x_{i+1}$ is another strong path in G with initial vertex x. By the choice of P, $S(P) \geq S(P')$. Note that $\mu(x_i x_{i+1}) \geq \mu(x_i x_m)$. Summing this over all $i \in S$, we obtain,

$$\sum_{i \in X} \mu(x_i x_{i+1}) \geq \sum_{i \in X} \mu(x_i x_m) = d_s(x_m) \geq t.$$

Let $C = x_j, x_m, x_{m-1}, \ldots, x_{j+2}, x_{j+1}$. Clearly C is a fuzzy cycle and the size of C is

$$\begin{aligned}
S(C) &= \sum_{i=j}^{m-1} \mu(x_i x_{i+1}) + \mu(x_m x_j) \\
&\geq \sum_{i \in X} \mu(x_i x_{i+1}) + \mu(x_m x_j) \\
&\geq t + \mu(x_m x_j) \\
&> t
\end{aligned}$$

Hence, C is a fuzzy cycle in G of size more than t. ■

Next we look at the fuzzy generalization of a celebrated result by Erdös and Gallai [49]. Let H be a fuzzy subgraph of G. For notational convenience, we let $d_s^H(x)$ and $N_s^H(x)$ denote the strong degree and strong neighbors set, respectively, of a vertex x.

Theorem 8.1.22 *Let $G = (\sigma, \mu)$ be a 2^*-connected fuzzy graph and t be a positive real number. Let x and y be distinct vertices of G. If $d_s(v) \geq t$ for every $v \in \sigma^* \backslash \{x, y\}$, where xy is not a fuzzy bridge, then G contains an $x - y$ strongest path of size at least t.*

Proof We use induction on $|\sigma^*|$. Let G be a 2^*-connected fuzzy graph with $|\sigma^*| = 3$ and let $\sigma^* = \{a, b, c\}$. Without loss of generality, we assume that $\mu(ab) = \mu(bc) = h_1$ and $\mu(ac) = h_0 \geq h_1$. In this case, t always belongs to $(0, 2h_1)$. For every pair of vertices, there exists a strongest path with size $h_0 + h_1 \geq t$. Suppose $|\sigma^*| \geq 4$ and the theorem is true for all fuzzy graphs on $|\sigma^*| - 1$ vertices. Let $H = (\sigma', \mu')$ where $\sigma'^* = \sigma^* \backslash \{y\}$ and $\mu'(uv) = \mu(uv)$ for $u, v \in \sigma'^*$.
Case 1. H is 2^*-connected.

Let $\epsilon = \max\{\mu(yu) : u \in N_s^H(y)\}$. Choose $w \in N_s(y) \backslash \{x\}$ (we can choose such a vertex since $|N_s^H(y)| \geq 2$) such that $\mu(yw) = \epsilon$. Then for every vertex $u \in \sigma'^* \backslash \{x\}$,

$$d_s^H(u) = d_s^G(u) - \mu'(yu) \geq t - \epsilon.$$

Let $t' = t - \epsilon$. By induction hypothesis, there exists an $x - w$ strongest path Q in H of size at least t'. Now set $P = Q \cup yw$. Then P is an $x - y$ strongest path in G of size at least t (See Fig. 8.6).

Case 2. H is not 2^*-connected.

Refer Fig. 8.7. In this case, H may contain fuzzy cut vertices. Now choose a block $B = (\tau, \nu)$ of H with exactly one fuzzy cutvertex of H such that $x \notin \tau^*$. Let z be the unique fuzzy cut vertex of H contained in B. Let $B' = (\tau', \nu')$ be the fuzzy subgraph of G induced by $\tau \cup \{y\}$.

If $y \in N_s(z)$, then B' is 2^*-connected and for all $w \in \tau' \backslash \{v, z\}$,

$$d_s^{B'}(u) = d_s^G(u) \geq t.$$

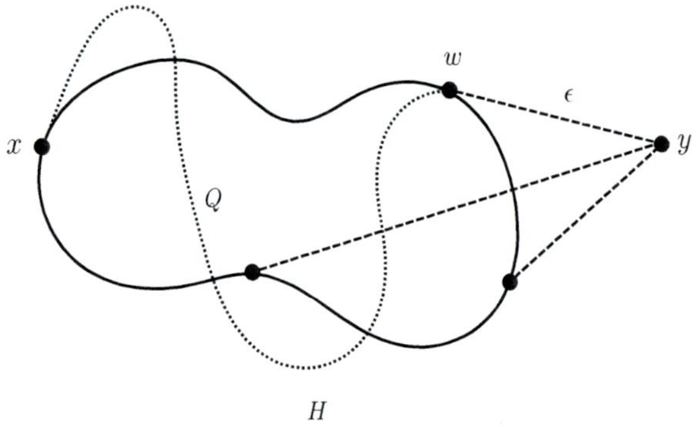

Fig. 8.6 Case 1 of the proof Theorem 8.1.22

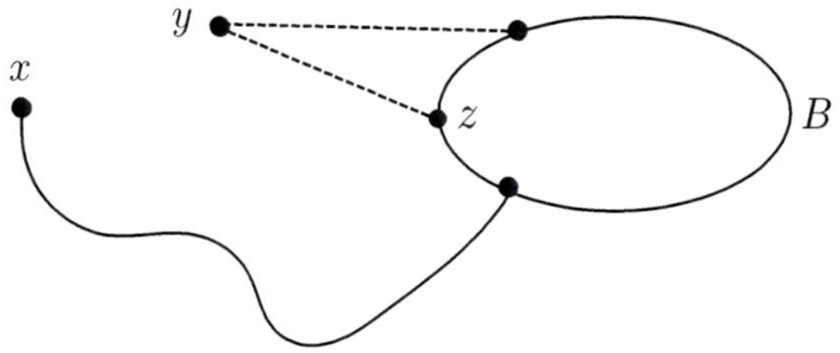

Fig. 8.7 Case 2 of the proof of Theorem 8.1.21

By the induction hypothesis, there is a strongest $y - z$ path in G of size at least t. Extending P in H from z to x, we obtain a strongest (x, y)-path in G of size at least t.

If $y \notin N_s(z)$, We add an edge yz to ν' setting $\nu'(yz) = CONN_G(y, z)$. By the induction hypothesis, there exists a strongest $y - z$ path P of size at least t. If $t > CONN_G(y, z)$, then P cannot be the edge ya, since $\mu(yz) = CONN_G(y, z)$. If $t = CONN_G(y, z)$, then we can choose P such that $P \neq yz$, because we need $S(P) \geq t$. Now we get a strongest $y - z$ path P with size at least t. Extending P as before, the proof will be complete. ■

Corollary 8.1.23 *Let $G = (\sigma, \mu)$ be a 2^*-connected fuzzy graph and t, a positive real number. If $d_s(v) \geq t$ for every $v \in \sigma^* \backslash \{x\}$, then G contains a strongest path of size at least t from x to any other vertex y of G, where xy is not an α- strong edge.*

8.2 Spanning Connectivity in Fuzzy Graphs

We now discuss a special type of container namely, a spanning container in a fuzzy graph. In graph theory, a container $C(u, v)$ of G is called a spanning container if $C(u, v)$ contains all the vertices of G [81]. Similarly, a spanning container in a fuzzy graph can be defined as follows.

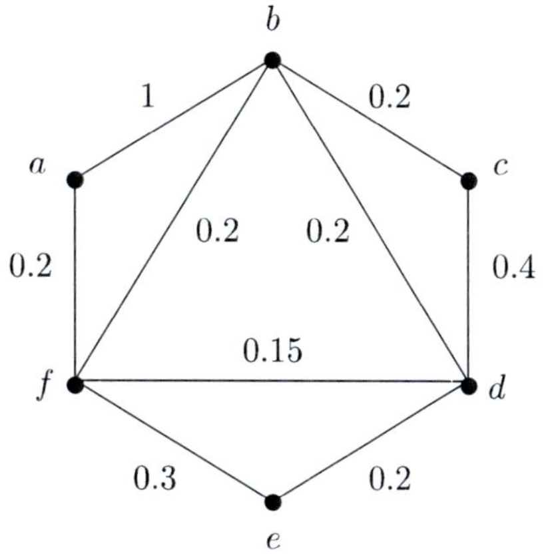

Fig. 8.8 A spanning 2-connected fuzzy graph with a (b, f)-spanning 3-fuzzy container

Definition 8.2.1 A fuzzy container $C_G(u, v) = (\tau, \nu)$ of $G = (\sigma, \mu)$ is said to be a *(u,v)-spanning fuzzy container* if $\tau^* = \sigma^*$.

Definition 8.2.2 A fuzzy graph $G = (\sigma, \mu)$ is said to be *spanning connected* if there exists a spanning fuzzy container between every pair of vertices $u, v \in \sigma^*$, except when uv is a fuzzy bridge. G is said to be *spanning k-connected* if for every pair of vertices u, v, there exists a spanning k-fuzzy container in G, except when uv is a fuzzy bridge.

Example 8.2.3 Consider the fuzzy graph $G = (\sigma, \mu)$ in Fig. 8.8 with $\sigma^* = \{a, b, c, d, e, f\}$, $\mu(ab) = 1$, $\mu(af) = \mu(df) = \mu(bd) = \mu(de) = \mu(bc) = 0.2$, $\mu(cd) = 0.4$, $\mu(df) = 0.15$ and $\mu(ef) = 0.3$. Here $C_G(b, d)$ and $C_G(b, f)$ are spanning 3-fuzzy containers of G. Also, $C_G^2(a, f)$, $C_G^2(f, c)$, $C_G^2(f, d)$, $C_G^2(a, c)$, $C_G^2(a, d)$, $C_G^2(a, e)$, $C_G^2(b, c)$, $C_G^2(b, e)$, $C_G^2(c, e)$ and $C_G^2(d, e)$ are spanning 2-fuzzy containers of G. Thus, G is spanning connected. Moreover, G is spanning 2-connected.

If $G = (\sigma, \mu)$ is a complete fuzzy graph with $|\sigma^*| = n > 2$ and $\sigma(v_1) \leq \sigma(v_2) \leq \cdots \leq \sigma(v_{n-1}) \leq \sigma(v_n)$, then by Theorem 8.1.8, we can find an $(n-1)$-fuzzy container between v_1 and v_i for $i = 2, 3, \ldots, n$ in G. Clearly, $C_G(v_1, v_i)$ is a spanning $(n-1)$-fuzzy container in G, for $i = 2, 3, \ldots, n$. Consider the complete fuzzy graph $G = (\sigma, \mu)$ on six vertices with $\sigma(a) = \sigma(b) = \sigma(c) = 0.1 < \sigma(d) = 0.11 < \sigma(e) = 0.3 < \sigma(f) = 0.4$ in Fig. 8.9a (0.15 is the membership value of each edge that are not mentioned in the figure). We observe that the fuzzy container $C_G(d, e)$ contains only three vertices; that is, d, e and f, and hence there does not exist a (d, e)-spanning fuzzy container in G. Hence, G is not spanning connected.

For the complete fuzzy graph $H = (\tau, \nu)$ with $\tau(a) = \tau(b) = \tau(c) = \tau(d) = 0.1 < \tau(e) = 0.3 < \tau(f) = 0.4$ in Fig. 8.9b (0.15 is the membership value of each edge that are not mentioned in the figure). We have that for every pair of vertices other than the fuzzy bridge ef, there exist a spanning fuzzy container in H. Thus, H is spanning connected. We summarize these results in the following result.

Theorem 8.2.4 *A complete fuzzy graph $G = (\sigma, \mu)$ is spanning connected if and only if $\sigma(v_i) = \sigma(v_j)$ for $i < j$, $j = 2, 3, \ldots, n-2$.*

For a spanning 2-connected fuzzy graph G, there always exists a strongest strong cycle containing every pair of vertices except the end vertices of fuzzy bridges. It can be seen that the only spanning 1-connected fuzzy tree is the fuzzy tree whose underlying graph is K_2.

Theorem 8.2.5 *A fuzzy tree $G = (\sigma, \mu)$ with $|\sigma^*| \geq 3$ is not spanning connected.*

Proof Let G be a fuzzy tree with $|\sigma^*| \geq 3$ and let F be the maximal spanning tree of G. Then there exists exactly one strongest path between any two vertices of σ^* in F. Let u be an internal vertex of F and v be any other vertex of F. Then clearly there does not exist a spanning 1-container between u and v. Thus, there does not exist a spanning container between every pair of vertices for such a fuzzy tree. Hence the theorem holds. ■

Theorem 8.2.6 *Let $G = (\sigma, \mu)$ be a cycle with $|\sigma^*| \geq 3$. Then G is locamin if and only if G is spanning 2-connected.*

Proof Suppose that G is spanning 2-connected. If possible assume that G is not locamin. Let xy be a weakest edge of G. Then there exists a vertex u in σ^* such that $\mu(vu)$ and $\mu(uw)$ are greater than $\mu(xy)$, where $v, w \in N_s(u)$. Let $P_1 = v - u - w$ and $P_2 = v - x - y - w$ (v may equal to x and w may equal to y). Then $s(P_1)$ will be greater than $s(P_2)$ in G. Thus $C_G^1(v, w) = \{P_1\}$. This implies that there exists only one 1-fuzzy container between v and w in G, a contradiction to the assumption that G is spanning 2-connected.

Conversely, assume that G is locamin cycle. Then every vertex of the cycle lies on a weakest edge. Hence there exists a 2-fuzzy container between every pair u, v of vertices except when uv is a fuzzy bridge containing all the vertices of G. ■

A $u - v$ path P in a fuzzy graph G is a strongest $u - v$ path if $s(P) = CONN_G(u, v)$. In general, a strongest path need not be a strong path and a strong path need not be a strongest path. But in a block, there always exists a strongest strong path between every two vertices of G. Hence if G is spanning k-connected then each path in $C_G^k(u, v)$ is a strongest strong path, for every pair of vertices $u, v \in \sigma^*$.

Theorem 8.2.7 *Let $G = (\sigma, \mu)$ be a fuzzy tree which is a cycle. Let u and v be any two vertices of G. Then $C_G(u, v)$ is a (u, v)-spanning 1-fuzzy container in G if and only if u and v are the fuzzy end vertices of the unique maximum spanning tree F of G.*

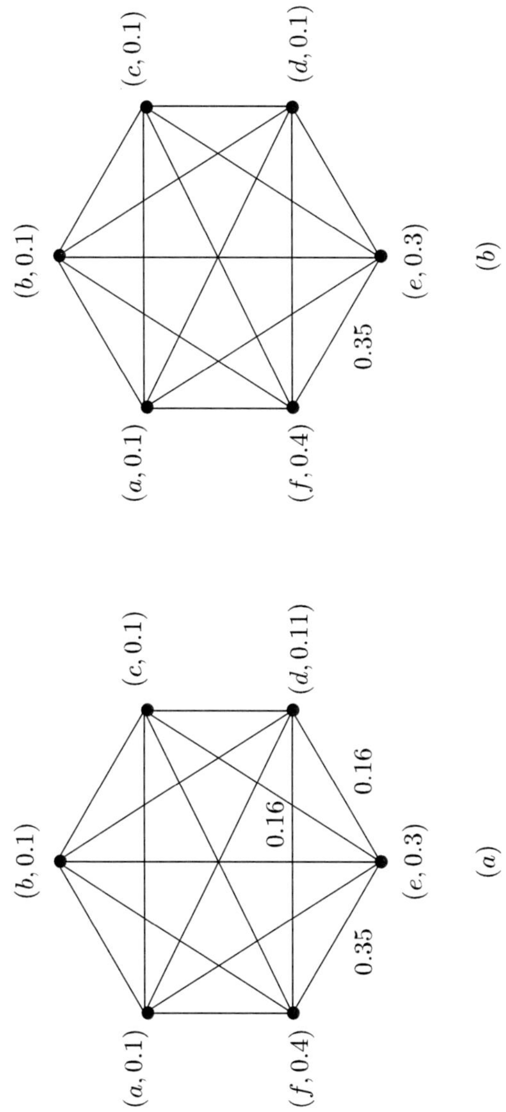

Fig. 8.9 **a** A CFG that is not spanning connected. **b** A spanning connected CFG

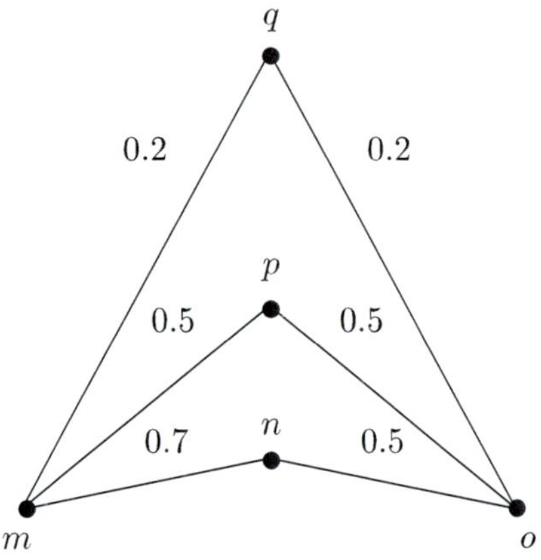

Fig. 8.10 A fuzzy graph which is a block, but not spanning connected

Proof Let $C_G^1(u, v)$ be a (u, v)-spanning 1-fuzzy container in G. If possible suppose that, either u or v or both is not an end vertex of the unique maximum spanning tree F. Without loss of generality, assume that u is not a fuzzy end vertex of F. The vertex u has at least two strong neighbors say, x and y. Hence there exists a strongest $u - v$ path P in F such that either x or y does not belong to the vertex set of P. Thus $P \cup C_G^1(u, v)$ is a 2-fuzzy container in G, which gives rise to a fuzzy cycle. However, is a contradiction to the fact that G is a fuzzy tree.

Conversely, suppose that u and v are the fuzzy end vertices of the unique maximum spanning tree F. Note that, G is a cycle and the $u - v$ path in F is the only strongest $u - v$ path in G. Hence $C_G^1(u, v)$ is the only (u, v)-spanning 1-fuzzy container in G. ■

Proposition 8.2.8 *Let $G = (\sigma, \mu)$ be a fuzzy cycle which is a block. Then G is spanning 2-connected.*

Proof Let G be a block. By Theorem 2.7.9 of [90], for any two vertices x, y of G, there exists a cycle C containing the vertices x and y which is formed by two strongest strong $x - y$ paths. Clearly, the cycle C is G itself. Thus it contains all the vertices of G. Hence for any given pair of vertices u, v, there exists two strongest paths between u and v which are internally disjoint and contains all vertices. Thus G is spanning 2-connected. ■

Corollary 8.2.9 *A fuzzy graph $G = (\sigma, \mu)$ is spanning k-connected if and only if every vertex of G has at least k strong neighbors.*

Figure 8.10 is an example of a fuzzy graph which is a block, but not spanning connected.

Theorem 8.2.10 *A spanning 2-connected fuzzy graph is a block.*

Proof Let $G = (\sigma, \mu)$ be a fuzzy graph satisfying the condition of the theorem. Choose an arbitrary vertex $z \in \sigma^*$. Then we can find a pair of vertices u, v such that $u \neq z \neq v$. Let $C_G(u, v) = \{P_1, P_2, \ldots, P_k\}$ for $k \geq 2$. That is, there exist k internally disjoint strongest $u - v$ paths in G. Since G is spanning 2-connected, at most one of these paths can contain the vertex z. Hence z cannot be a fuzzy cutvertex of G. Since z is arbitrary, it follows that G is a block. ■

The converse of the above result is not true in general as seen from the following example.

Example 8.2.11 Let $G = (\sigma, \mu)$ be with $\sigma^* = \{m, n, o, p, q\}$, $\mu(no) = \mu(op) = \mu(pm) = 0.5$, $\mu(qm) = \mu(oq) = 0.5$ and $\mu(mn) = 0.7$ (Fig. 8.10). Here q does not belong to the vertex set of $C_G(p, m)$, $C_G(p, o)$ and $C_G(n, o)$. Hence, G is not a spanning connected, but G is a block.

8.3 Algorithms and Application to Human Trafficking

Two major algorithms will be discussed in this section. The first one is to find the maximum capacity of a container between a given pair of vertices and the second to find maximum and minimum container capacities of a given fuzzy graph. Also, a an application on human trafficking is given at the end.

Algorithm 8.3.1 Given a connected fuzzy graph $G = (\sigma, \mu)$ with $|\sigma^*| = n$. This algorithm is to find a fuzzy container between u and v with maximum capacity in G.

1. Find the strength of connectedness $CONN_G(u, v)$ between u and v using the algorithm in [17].
2. Let u be such that $|N_s(u)| \leq |N_s(v)|$. Let $N(u) = \{u_1, u_2, \ldots, u_r\}$.
3. Initialize $i = 1, k = 1$ and $X_i = \{\}$.
4. If $u_i = v$, find $\mu(uv)$.

    4.1. If $\mu(uv) = CONN_G(u, v)$. Then take $L_k = uv$. Add L_k to the set X_i and go to step 4 by setting $i = i + 1$ and $k = k + 1$.
    4.2. Else, go to step 4 by setting $i = i + 1$.

5. Else if there does not exists a $u - v$ path with u_i as internal vertex, Then go to step 4 by setting $i = i + 1$.
6. If there exist l number of $u - v$ paths, say $\{P_{ij}\}_{j=1}^{l}$, with u_i as internal vertex, then,

    6.1. Let $j = 1$. If P_{ij} is internally disjoint from L_p, for all $p < k$ and $s(P_{ij}) = CONN_G(u, v)$, then take $P_{ij} = L_k$. Add L_k to the set X_i and go to step 4 by setting $i = i + 1$ and $k = k + 1$.

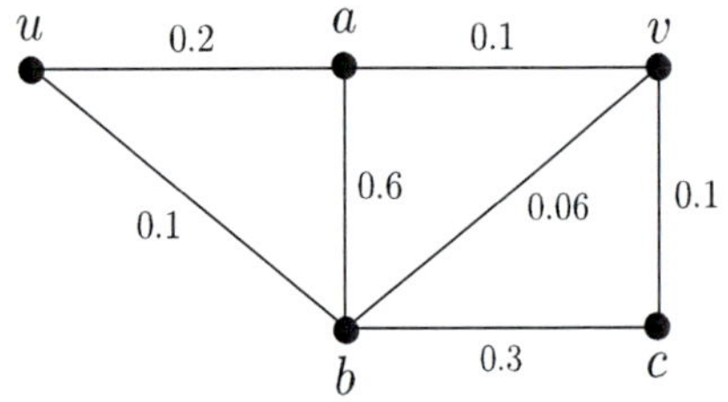

Fig. 8.11 Fuzzy graph with a (u, v)-fuzzy container having capacity 0.2

6.2. Else, if $j < l$, set $j = j + 1$ and go to 6.1 else, go to step 4 by setting $i = i + 1$.

7. If every vertex in $N(u)$ has been marked visited, stop.
8. X_i is a fuzzy container between u and v in G.
9. Repeat the algorithm by reordering vertices in $N(u)$ in such a way that u_1 is replaced by u_i and correspondingly we obtain the set X_i for $i = 2, 3, \ldots, r$. Hence $X_1, X_2, \ldots, X_{r-1}$ and X_r are different fuzzy containers between u and v. Take $\epsilon = \max\{|X_i| : i = 1, 2, \ldots, r\}$. Then $Cap(C_G(u, v)) = \epsilon CONN_G(u, v)$ is maximum capacity of a (u, v)-fuzzy container in G.

Illustration of Algorithm 8.3.1.

Consider the fuzzy graph $G = (\sigma, \mu)$ with $\sigma^* = \{a, b, c, u, v\}$, $\mu(av) = \mu(ub) = \mu(cv) = 0.1$, $\mu(ua) = 0.2$, $\mu(bc) = 0.3$ $\mu(ab) = 0.6$ and $\mu(bv) = 0.06$ (Fig. 8.11). With the help of the algorithm proposed by Banarjee [17], we obtain $CONN_G(u, v) = 0.1$. From the figure it is clear that $|N(u)| = 2 < |N(v)| = 3$. Thus we use the vertex u to start the algorithm. $N(u) = \{a, b\}$, let $u_1 = a$ and $u_2 = b$. We start with u_1, and there exist three $u - v$ paths in G. There are three $u - v$ paths with a as an internal vertex. Let $P_{11} = u - a - v$, $P_{12} = u - a - b - v$ and $P_{13} = u - a - b - c - v$. Beginning with P_{11}, $s(P_{11}) = 0.1$ which is equal to $CONN_G(u, v)$. Take $L_1 = P_{11}$ and added to X_1. See Fig. 8.11.

Hence go to step 4 by setting $i = 2$ and $k = 2$. Here $u_2 = b$ and there are three $u - v$ paths with b as internal vertex. Let $P_{21} = u - b - v$, $P_{22} = u - b - a - v$ and $P_{23} = u - b - c - v$. $s(P_{21}) = 0.06$ which is not equal to $CONN_G(u, v)$. Now go to step 6 by setting $j = 2$, that is P_{22}. $s(P_{22})$ is 0.1 which is equal to $CONN_G(u, v)$. Take $L_2 = P_{22}$ and added to X_1.

Every vertex in $N(u)$ has been marked visited. $X_1 = \{P_{11}, P_{11}\}$ Now let $N(u) = \{u_1 = b, u_2 = a\}$. We start with u_1. There are three $u - v$ paths with b as internal vertex. As above, let $P_{11} = u - b - v$, $P_{12} = u - b - a - v$ and $P_{13} = u - b - c - v$. $s(P_{11}) = 0.06$ which is not equal to $CONN_G(u, v)$. Now go to step 6.1 by setting $j = 2$, that is P_{12}. $s(P_{12})$ is 0.1 which is equal to $CONN_G(u, v)$. Take $L_1 = P_{12}$ and added to X_2

Hence go to step 4 by setting $i = 2$. There exists three $u - v$ paths in G. There are three $u - v$ paths with a as an internal vertex. Let $P_{21} = u - a - v$, $P_{22} = u - a -$

$b - v$ and $P_{23} = u - a - b - c - v$. But, P_{21}, P_{22} and P_{23} shares a vertex a with L_1. Thus we stop. Since, every vertex in $N(u)$ has been marked visited, $X_2 = \{P_{12}\}$.

Now moving to step 9, $|X_1| = 2$ and $|X_2| = 1$, $\epsilon = 2$. Consequently, $Cap(C_G(u, v)) = \epsilon CONN_G(u, v) = 2 \times 0.1 = 0.2$. Hence maximum capacity of (u, v)-fuzzy container in G is 0.2.

Algorithm 8.3.2 Given a fuzzy graph $G = (\sigma, \mu)$ with $|\sigma^*| = n$. This algorithm is to find $minCap$ and $maxCap$ of G.

1. For each $u, v \in \sigma^*$, find the strength of connectedness $CONN_G(u, v)$ between u and v using the algorithm in [17].
2. Find (u, v)-fuzzy container with maximum capacity using the Algorithm 8.3.1.
3. Find $\xi = \min\{ Cap(C_G(u, v)) = u, v \in \sigma^*, uv$ is not a fuzzy bridge $\}$.
4. Find $\eta = \max\{ Cap(C_G(u, v)) = u, v \in \sigma^*, uv$ is not a fuzzy bridge $\}$.
5. $C_G(a, b) = \xi$ and $C_G(c, d) = \eta$ for some $a, b, c, d \in \sigma^*$, are $minCap(g)$ and $maxCap(G)$ respectively.

Illustration of Algorithm 8.3.2

Consider the previous example. With the help of the Algorithm 8.3.1, we obtain $Cap(C_G(m, q)) = Cap(C_G(q, o)) = 0.1$, $Cap(C_G(m, p)) = Cap(C_G(p, o)) = Cap(C_G(n, o)) = 0.4$, $Cap(C_G(m, n)) = 0.6$. By step 3 and 4 of Algorithm 5.2, $\xi = 0.2$ and $\eta = 0.4$. Hence $C_G(m, n)$ is a fuzzy container which gives $minCap(G) = 0.2$ and $C_G(m, q) = 0.4$ is a fuzzy container which gives $maxCap(G)$.

Application to Human Trafficking

Mordeson and Mathew [109] used fuzzy graph models for the determination of illegal emigration routes related with human trafficking. Here the model is in the form of a directed fuzzy graph. Study of containers in directed fuzzy graphs can be studied analogously to undirected graphs. The data of flow from country to country is taken from [109]. Consider the flow graph G on the vertex set $\sigma^* = \{a, b, c, d, e, f, g, h, i, j\}$, obtained in [109]. We find the strength of connectedness between countries and fuzzy containers between them.

We observe from Table 8.1 that the normalized data flow of emigrants between certain vulnerable regions of the globe. It can be expressed with a directed fuzzy graph model G as in Fig. 8.12.

If $G = (V, \sigma, \mu)$ is a DFN with $\sigma^* = \{v_1, v_2, \ldots, v_n\}$, , then the container matrix, denoted by $C_f(\overrightarrow{G})$ is an $n \times n$ matrix where $C_f(\overrightarrow{G})[i, j] = Cap(C_G(v_i, v_j))$, if $i \neq j$ and 0 if $i = j$. By using Algorithm 8.3.1, we obtain the entries $C_f(\overrightarrow{G})[i, j]$, for $i \neq j$. The following is the container matrix $C_f(\overrightarrow{G})$ of the fuzzy graph in Fig. 8.12.

Table 8.1 Illegal emigration between 10 different locations

	WCE(a)	WSE(b)	CE & Bal(c)	EE & CAs(d)	NA & CA & Car(e)	SA(f)	EAs & Pac(g)	SAs(h)	SSA(i)	Mid East(j)
WC Eur(a)	0.62	0.13								
WS Eur(b)		0.16								
C Eur & Bal(c)		0.27	0.79		0.05					
E Eur & C Asia(d)	0.04	0.04	0.05	0.99						0.06
N Am & C Am & Car(e)	0.08	✓	✓		0.59	0.04				
S Am(f)		0.07			0.03	0.94				
E Asia & Pac(g)	0.07	0.07	✓		0.25	0.01	0.97			0.33
S Asia(h)		✓			0.07			0.96		0.18
S S Afr(i)		0.16						1.0		0.10
Mid East(j)										0.31

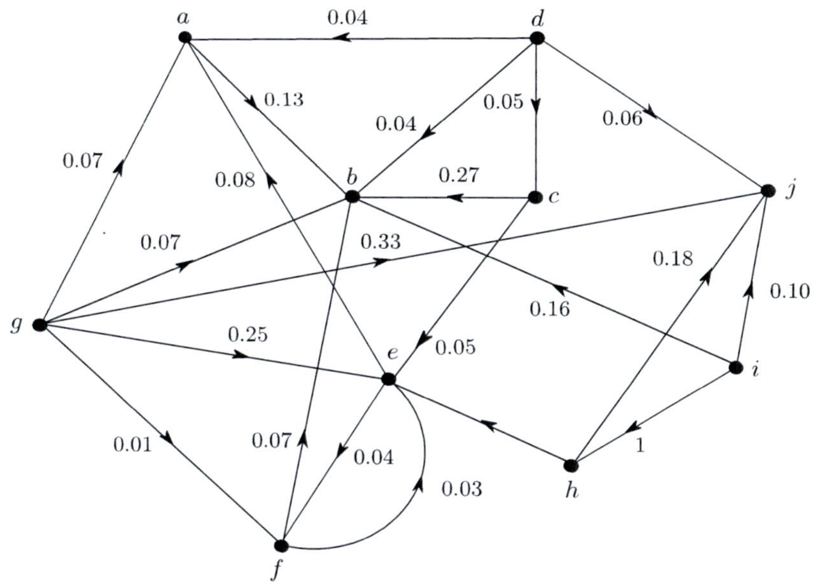

Fig. 8.12 Flow graph between different parts of the globe

$$
C_f(\overrightarrow{G}) = \begin{array}{c} \\ a \\ b \\ c \\ d \\ e \\ f \\ g \\ h \\ i \\ j \end{array}
\begin{array}{c}
\begin{array}{cccccccccc} a & b & c & d & e & f & g & h & i & j \end{array} \\
\left[\begin{array}{cccccccccc}
0 & 0.13 & 0 & 0 & 0 & 0 & 0 & 0 & 0 & 0 \\
0 & 0 & 0 & 0 & 0 & 0 & 0 & 0 & 0 & 0 \\
0.05 & 0.27 & 0 & 0 & 0.05 & 0.04 & 0 & 0 & 0 & 0 \\
0.05 & 0.05 & 0.05 & 0 & 0.05 & 0.04 & 0 & 0 & 0 & 0.06 \\
0.08 & 0.08 & 0 & 0 & 0 & 0.04 & 0 & 0 & 0 & 0 \\
0.03 & 0.07 & 0 & 0 & 0.03 & 0 & 0 & 0 & 0 & 0 \\
0.08 & 0.07 & 0 & 0 & 0.25 & 0.04 & 0 & 0 & 0 & 0.33 \\
0.07 & 0.07 & 0 & 0 & 0.07 & 0.04 & 0 & 0 & 0 & 0.18 \\
0.07 & 0.16 & 0 & 0 & 0.07 & 0.04 & 0 & 1 & 0 & 0.18 \\
0 & 0 & 0 & 0 & 0 & 0 & 0 & 0 & 0 & 0
\end{array}\right]
\end{array}
$$

From the figure, we can see that the directed fuzzy graph is a directed fuzzy tree. Using Algorithm 8.3.1 we obtain entries of the container matrix $C_f(\overrightarrow{G})$. With the help of Algorithm 8.3.2, $minCap(G) = Cap(C_G(f, e)) = 0.03$ and $maxCap(G) = Cap(C_G(g, j)) = 0.33$, and obtain minimum and maximum fuzzy container as $C_G(f, e)$ and $C_G(g, j)$ with strength 0.03 and 0.33 respectively. Being a fuzzy tree, fuzzy vertex connectivity is $\min\{\mu(uv) : uv$ is a strong edge in$G\} = 0.03$, which is same as strength of a minimum fuzzy container of G.

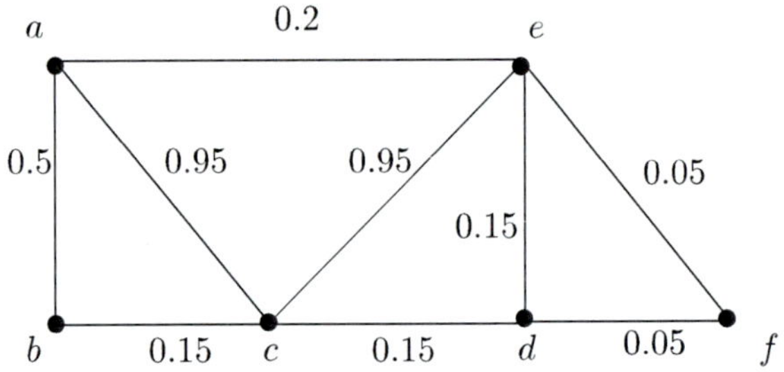

Fig. 8.13 Fuzzy graph in Example 8.4.2

8.4 Cycles in Fuzzy Graphs

In the mathematical analysis of networks, there are two primary problems to be addressed: first to identify paths with maximum capacity and the second to find paths with maximum data flow. It is always desirable that there exist several maximum flow paths between pairs of nodes. It will help in avoiding traffic delays and bottlenecks. Also, in quality of service problems, multiple paths can help in providing extra service to customers using different routes. Fuzzy graph structures with two or more different strongest paths between every pairs of nodes were proposed and studied by Mathew and Sunitha [96] and Mordeson and Mathew [90] in recent years. Different types of cycles also were studied. We focus paths and cycles in this section. The formal definitions of a heavy path and a heavy cycle are as follows.

Definition 8.4.1 A path of maximum size is said to be a **heavy path**. A cycle $C = (\tau, \nu)$ in G is said to be a **heavy cycle** if $\sum xy \in \nu^{*}\nu(xy)$ is maximum among all cycles in G. Such a cycle C is a **heavy fuzzy cycle** if C is also a fuzzy cycle in G.

Example 8.4.2 Consider the fuzzy graph in Fig. 8.13. In this fuzzy graph, we have, $\mu(ab) = 0.5$, $\mu(bc) = \mu(cd) = \mu(ed) = 0.15$, $\mu(df) = \mu(ef) = 0.05$ $\mu(ac) = \mu(ce) = 0.5$ and $\mu(ae) = 0.2$. Here, the path $P : a - c - e - d - f$ is heavy with size 2.1 and cycle $C : a - c - e - a$ is a heavy cycle with size 2.1. Clearly $a - b - c - d - f - e - a$ and $e - c - d - f - e$ are heavy fuzzy cycles with sizes 1.1 and 1.2 respectively.

In any fuzzy graph, a heavy path is not necessarily a strong path or a strongest path. A heavy fuzzy cycle is always a heavy cycle, but the converse is not true. In complete fuzzy graphs, heavy cycles and heavy fuzzy cycles coincide as all its edges are strong.

Theorem 8.4.3 *If G is a block and P is a heavy path in G from x to y, then there is a cycle C in G with the property that either $S(C) > S(P)$ or $S(C) \geq 2CONN_G(x, y) + 2s(P)$.*

Proof Suppose $P = (\tau, \nu)$ with $\tau^* = \{u_0, u_1, \ldots, u_k\}$ is a heavy path from x to y in G. Assume $u_0 < u_1 < \cdots < u_k$, where ' $<$ ' denotes the precedence on P. Let $A = \{u_i : xu_i \in \mu^*\}$ and $B = \{u_i : u_{i-1}y \in \mu^*\}$.

Since P is a heavy path, $N(x, G) \cup N(y, G) \subseteq \tau^*$. We have that

$$\mu(xu_i) \geq CONN_G(x, y) \quad \forall u_i \in A \tag{8.4.1}$$

$$\mu(u_{i-1}y) \geq CONN_G(x, y) \quad \forall u_i \in B \tag{8.4.2}$$

$$\mu(xu_i) \leq \sum j = 1i\mu(u_{j-1}u_j) \quad \forall u_i \in A \tag{8.4.3}$$

$$\mu(u_{i-1}y) \leq \sum j = il\mu(u_{j-1}u_j) \quad \forall u_i \in B \tag{8.4.4}$$

Suppose that A and B are not disjoint.

If $y \in A \cap B$, then $\mu(xy) > 0$. Let $C = (\tau'^*, \nu'^*)$ be the fuzzy subgraph such that $\tau'^* = \tau^*$ and $\nu'^* = \nu^* \cup yx$. Then C is a cycle and has size $S(C) = S(P) + \mu(xy) > S(P)$.

Otherwise, for $u_i \in A \cap B$, define

$$C_j = u_0u_1 \cdots u_{j-1}u_ku_{k-1} \cdots u_ju_0$$

The cycle C_j covers the edges $u_{j-1}u_j \notin A \cap B$ and the edges xu_j and $u_{j-1}y$, $u_j \in A \cap B$. Therefore,

$$\begin{aligned} S(C_j) &= \sum_{i \neq j}\{\mu(u_{i-1}u_i) + \mu(xu_j) + \mu(u_{j-1}y) \\ &= \sum j = 1i\{\mu(u_{i-1}u_i) + \sum j = il\{\mu(u_{i-1}u_i) - \mu(u_{i-1}u_i) + \mu(xu_j) + \mu(u_{j-1}y) \end{aligned}$$

$$\begin{aligned} &\geq \sum j = 1i\{\mu(u_{i-1}u_i) + \sum j = il\{\mu(u_{i-1}u_i) + \mu(xu_j) + \mu(u_{j-1}y) \\ &\geq \mu(xu_j) + \mu(u_{j-1}y) + \mu(xu_j) + \mu(u_{j-1}y) \quad \text{(using (8.4.3) and (8.4.4))} \\ &\geq 4CONN_G(x, y) \quad \text{(using (8.4.1) and (8.4.2))} \\ &\geq 2CONN_G(x, y) + 2s(P). \end{aligned}$$

Suppose now that $A \cap B = \emptyset$.

We define a set $Q = \{Q_j : (\tau_j, \nu_j)\}_1^l$ of strongest paths which are disjoint internally in G (see Fig. 8.14) with the following properties.

1. $\tau^* \cap \tau_j^* = \{a_j, b_j\}$ $1 \leq j \leq l$;
2. $x = a_1 < a_2 < b_1 \leq a_3 < b_2 < a_4 < \cdots \leq a_m < b_{m-1} < b_m = y$.

Here a_j and $b_j$represent the end vertices of Q_j and $<$ denotes the precedence on the path P. We note that $S(Q_i) \geq s(P)$ for $i = 1, l$. The notation $P[a_i, b_i]$ is used to describe an $a_i - b_i$ path in P with origin a_i and terminus b_i. Since G is block, by

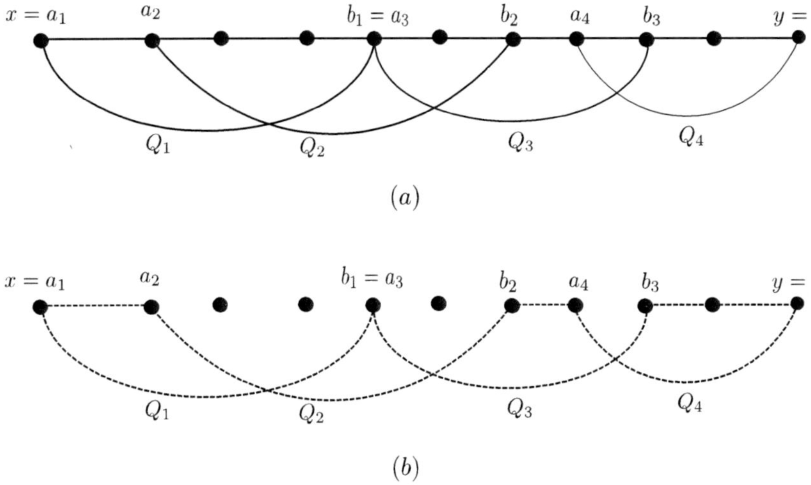

Fig. 8.14 Construction of cycle C in Case 2

Menger's theorem for fuzzy graphs, $|Q| \geq 2$. Suppose U_j represents the collection of internal vertices in the segment $[a_{j+l}, b_j]$, $1 < j \leq l-1$ of P. Choose Q such that, l and $|\bigcup_{j=1}^{l-1} U_j|$ are as small as possible.

Now construct a cycle $C = (\sigma', \mu')$ with, $\sigma'^* = (\tau^* \setminus \{U_1 \cup U_2 \cup \cdots \cup U_{l-1}\}) \cup \tau_1^* \cup \tau_2^* \cdots \cup \tau_l^*$ and

$$\mu'(uv) = \begin{cases} \nu(uv) & \text{if } u, v \in \tau^* \setminus \{U_1 \cup U_2 \cup \cdots \cup U_{l-1}\} \\ \nu_j(uv) & \text{if } u, v \in \tau_j^* \end{cases}$$

Here, Q_1 and Q_l are the edges a_1b_1 and a_lb_l respectively. Hence $S(Q_1) = \mu(a_1b_1) = CONN_G(a_1, b_1)$ and $S(Q_l) = \mu(a_lb_l) = CONN_G(a_l, b_l)$. Moreover we claim that C has the required property. By the choice of Q, with the assumption $A \cap B = \emptyset$,

$$\begin{aligned} S(C) &= \sum_{j=1}^{l} S(Q_j) + \sum_{j=1}^{n-1} \mu(b_j a_{j+2}) + S(P[a_1, a_2]) + S(P[b_{n-1}, b_n]) \\ &\geq \sum_{j=1}^{l} S(Q_j) + \sum_{j=1}^{n-1} \mu(b_j a_{j+2}) + 2CONN_G(x, y) \quad \text{(using (8.4.1) and (8.4.2))} \\ &\geq 2s(P) + \sum_{j=1}^{n-1} \mu(b_j a_{j+2}) + 2CONN_G(x, y) \\ &\geq 2s(P) + 2CONN_G(x, y) \end{aligned}$$

This completes the proof. ■

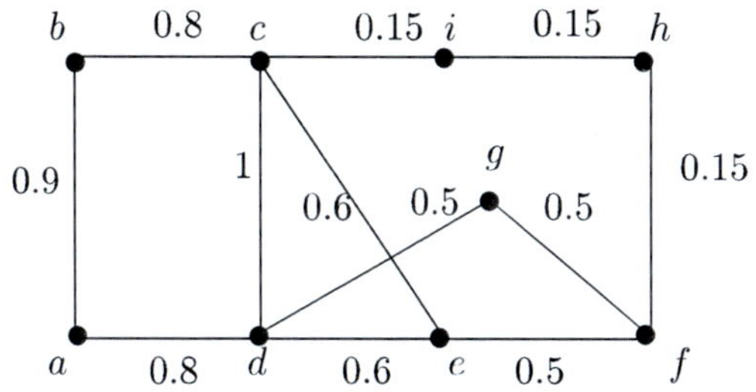

Fig. 8.15 Fuzzy graph with an $a - g$ heavy path

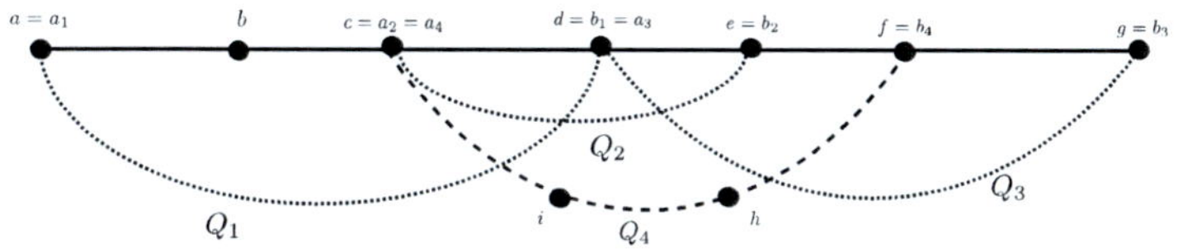

Fig. 8.16 Fuzzy graph G-illustration of special case of Theorem 8.4.3

Illustration

Consider the fuzzy graph in Fig. 8.15. The support has 9 vertices with $\mu(ab) = 0.9$ $\mu(ad) = \mu(bc) = 0.8$, $\mu(de) = \mu(ce) = 0.6$, $\mu(ef) = \mu(fg) = \mu(dg) = 0.5$ $\mu(fh) = \mu(hi) = \mu(ci) = 0.15$ and $\mu(cd) = 1$.

The path $P = a - b - c - d - e - f - g$ is a heavy path in G with $S(P) = 4$ whose strength is $s(P) = 0.5$ and $CONN_G(a, g) = 0.5$

As in the proof of the theorem, we get $A = \{u_i : au_i \in \mu^*\} = \{b, d\}$ and $B = \{u_i : u_{i-1}g \in \mu^*\} = \{c, g\}$. Hence, $A \cap B = \emptyset$. Further, we determine internally disjoint strongest paths in such a way that end vertices intersects P. These paths are $Q_1 = ad$, $Q_2 = dg$ $Q_3 = ce$ and $Q_4 = c - i - h - f$. Among the paths we choose two sets $Q = \{Q_1, Q_2, Q_3\}$ and $Q' = \{Q_1, Q_2, Q_4\}$ which satisfies conditions 1 and 2 of the proof. Vertex sets $\bigcup_{j=1}^{l-1} U_j$ with respect to Q and Q' are $\{e\}$ and empty set. We choose a set of internally disjoint strongest paths in such a way that, $|\bigcup_{j=1}^{l-1} U_j|$ is as small as possible. Thus we move with the set $Q = \{Q_1, Q_2, Q_3\}$. Now we construct C using Q as follows (See Fig. 8.16).

$$C : Q_1 \cup Q_3 \cup g - f - e \cup Q_2 \cup c - b - a.$$

Theorem 8.4.4 *Consider a heavy cycle $C = (\tau, \nu)$ in $G = (\sigma, \mu)$. If there exists a path P in $G - C$, connecting x and y in σ^*, with $|N(x, C)| \geq 1$, $|N(y, C)| \geq 1$, and $|N(x, C)| \cup |N(y, C)| \geq 2$, then $S(C) \geq 2 \min\{\mu(xa), \mu(yb)\} + 2S(P)$, where $a \in N(x, C)$ and $b \in N(y, C)$.*

Proof Choose a and b in such a way that $a \in N(x, C) \setminus N(y, C)$ and $b \in N(y, C) \setminus N(x, C)$. We define two paths say, $C_1[a, b]$ and $C_2[a, b]$ using C. Construct two new cycles $C_i, i = 1, 2$ using the segments $C_i[a, b]$. These two cycles are defined as follows,

$$C_i = C_i[a, b] \cup ax \cup by \cup P, i = 1, 2$$

Note that every edge of C is traversed exactly once in C_i. Edges ax and by are traversed twice and the path P is traversed two times. Then the average size of these 2 cycles is

$$\frac{1}{2}(S(C) + 2\mu(ax) + 2\mu(by) + 2S(P))$$

Since C is a heavy cycle, its size is not less than the average size. Thus,

$$\begin{aligned} S(C) &\geq \frac{1}{2}(S(C) + 2\mu(ax) + 2\mu(by) + 2S(P)) \\ &\geq 2\mu(ax) + 2\mu(by) + 2S(P)) \\ &\geq 2\min\{\mu(xa), \mu(yb)\} + 2S(P). \end{aligned}$$

■

A more general version of Theorem 8.4.4 is given below.

Theorem 8.4.5 *Suppose $C = (\tau, \nu)$ is a heavy cycle G and that there always exists a path $P : (\rho, \xi)$ in $G - C$, joining $x, y \in V(G)$, such that $|N(x, C)| \geq 1$, $|N(y, C)| \geq 1$, and $|N(x, C)| \cup |N(y, C)| \geq 2$. Define*

$$X = N(x, C) \setminus N(y, C), \quad Y = N(y, C) \setminus N(x, C) \text{ and } \quad Z = N(x, C) \cap N(y, C).$$

Then

$$S(C) \geq \begin{cases} 2d(x, Y) + 2\max\{d(x, Z) + S(P), d(y, Z)\} & \text{if } X = \emptyset \text{ and } |Z| = 1 \\ 2d(y, X) + 2\max\{d(y, Z) + S(P), d(x, Z)\} & \text{if } Y = \emptyset \text{ and } |Z| = 1 \\ 2d(x, X) + 2d(y, Y) + d(x, Z) + d(y, Y) + \max\{2, |Z|\}S(P) & \text{otherwise} \end{cases}$$

Proof Let $U = X \cup Y \cup Z$. Suppose $A = \{u_1, u_2, \ldots, u_n\}$, where u_i are in a given order around C. Two different cases arise.

Case 1: $X = \emptyset$ and $|Z| = 1$ or $Y = \emptyset$ and $|Z| = 1$.

We assume that $Z = \{u_1\}$ and $Y = \{u_2, u_3, \ldots, u_n\}$. Let $C' = (\tau', \nu')$, where $\tau'^* = \tau^* \cup \rho^*$ and $\nu'^* = (\nu^* \setminus \{u_1u_2\}) \cup \xi^* \cup \{u_1x, u_2y\}$ and $C'' = (\tau'', \nu'')$, where $\tau''^* = \tau^* \cup \rho^*$ and $\nu''^* = (\nu^* \setminus \{u_1u_2\}) \cup \xi^* \cup \{u_1y, u_2y\}$. Thus C' and C'' are two cycles in G with $S(C') \leq S(C)$ and $S(C'') \leq S(C)$. Hence,

$$S(C[u_1, u_2]) \geq \mu(u_1x) + S(P) + \mu(u_2y) \quad \text{and} \tag{8.4.5}$$

$$S(C[u_1, u_2]) \geq \mu(u_1y) + \mu(u_2y) \tag{8.4.6}$$

From (8.4.5) and (8.4.6),

$$S(C[u_1, u_2]) \geq \vee\{(\mu(u_1x) + S(P)), \mu(u_1y)\} + \mu(u_2y)$$

Similarly,

$$S(C[u_1, u_n]) \geq \vee\{(\mu(u_1x) + S(P)), \mu(u_1y)\} + \mu(u_ny)$$

For all $i \neq 1, n$, we have

$$S(C[u_i, u_{i+1}]) \geq \mu(u_iy) + \mu(u_{i+1}y)$$

$$\begin{aligned}
S(C) &\geq S(C[u_1, u_2]) + S(C[u_2, u_3]) + \cdots + S(C[u_{n-1}, u_n]) + S(C[u_1, u_n]) \\
&\geq \vee\{(\mu(u_1x) + S(P)), \mu(u_1y)\} + \mu(u_2y) + \vee\{(\mu(u_1x) + S(P)), \mu(u_1y)\} + \mu(u_ny) \\
&\quad + \sum i = 2n - 1\mu(u_iy) + \mu(u_{i+1}y) \\
&= 2(\vee\{(\mu(u_1x) + S(P)), \mu(u_1y)\}) + 2\sum i = 2n\mu(u_iy) \\
&= 2(\vee\{(d(x, Z) + S(P)), d(y, Z)\}) + 2d(y, Y)
\end{aligned}$$

With the same argument, for $Y = \emptyset$ and $|Z| = 1$, $S(C) \geq 2(\vee\{((d(x, Z) + S(P)), d(y, Z)\}) + 2d(x, X)$.

Case 2. $|Z| = 1$, $X \neq \emptyset$ and $Y \neq \emptyset$; $|Z| \geq 2$; and $|Z| = 0$.

Corresponding to each pair (u_i, u_{i+1}) of vertices, two new cycles are constructed from C by the replacement of the segment $C[u_i, u_{i+1}]$ with two new (u_i, u_{i+1})-paths where i varies from 1 to n and $n + 1$ is assumed as 1. These paths are defined as follows:

(i) If $u_i, u_{i+1} \in Z$, then the two paths are u_ixPyu_{i+1} and u_iyPxu_{i+1}.
(ii) If $u_i \in Z$ and $u_{i+1} \in X \cup Y$, then the paths are either u_iyPxu_{i+1} and u_ixu_{i+1} or u_ixPyu_{i+1} and u_iyu_{i+1}. If $u_{i+1} \in Z$ and $u_i \in X \cup Y$, the paths are defined similarly.
(iii) If $u_i \in X$ and $u_{i+1} \in Y$ or $u_i \in Y$ and $u_{i+1} \in X$, then the path is u_ixPyu_{i+1} or u_iyPxu_{i+1}.
(iv) If $u_i, u_{i+1} \in X$ and $u_i, u_{i+1} \in Y$, then the path is u_ixu_{i+1} or u_iyu_{i+1} (Fig. 8.17).

Using the above two cases, we have replaced the segment $C[u_i, u_{i+1}]$, with two paths and hence obtained two cycles. Note that in (iii) and (iv) these two cycles are identical. Because there exist k pairs of different vertices (u_i, u_{i+1}) where $i = 1, 2, \ldots, n$, we get $2n$ cycles. Here, we observe that,

(1) Each edge is repeated exactly $2(n - 1)$ times.
(2) An edge is traversed two times if it is from x or y to Z.
(3) An edge is traversed four times if it is from x to X.

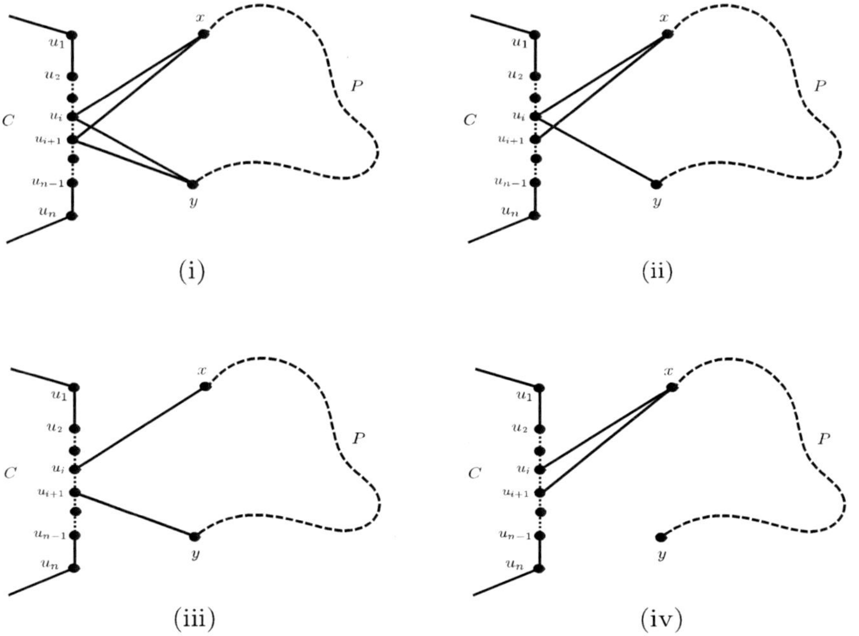

Fig. 8.17 Construction of cycles in Case 2

(4) An edge is traversed four times if it is from y to Y.

Let m denotes the number of times P is taken by these cycles. Then we can determine m with the assumption $|N(x, C)| \geq 1$ and $|N(y, C)| \geq 1$; cases (ii) and (iii), that,

$$m \geq \begin{cases} 2|Z| & \text{if } |Z| \geq 2 \\ 4 & \text{if } |Z| = 1, X \neq \emptyset \text{ and } Y \neq \emptyset \\ 4 & \text{if } |Z| = 0 \end{cases}$$

Altogether we can conclude that, $S(C) \geq 2d(x, X) + 2d(y, Y) + d(x, Z) + d(y, Z) + \max\{2, |Z|\}S(P)$ ■

Corollary 8.4.6 *Let C be a heavy cycle of $G = (\sigma, \mu)$. If there exists a path P in $G - C$, joining x and y, with the property that $|N(x, C)| \geq 1$, $|N(y, C)| \geq 1$, and $|N(x, C)| \cup |N(y, C)| \geq 2$, then $S(C) \geq 2 \min\{d(x, C), d(y, C)\} + 2\,S(P)$.*

Theorem 8.4.7 *Suppose G is connected and has no fuzzy cutvertices. Then there exist at least two vertices in common for any two heavy cycles in G.*

Proof Suppose that $C_1 : u_1u_2 \cdots u_ru_1$ and $C_2 : v_1v_2 \cdots v_sv_1$ are two heavy cycles in G. If C_1 and C_2 are disjoint, then by Menger's theorem, there always exist two

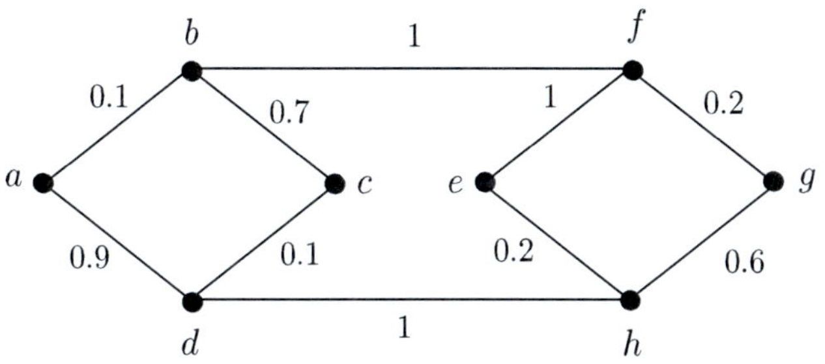

Fig. 8.18 A fuzzy graph with two heavy fuzzy cycles $a - b - c - d - a$ and $e - f - g - h - e$

paths strongest and internally disjoint, say, P_1 joining u_1 and v_1 and P_2 joining u_2 and v_2 connecting C_1 and C_2.

Let P_1 intersects C_1 as well as C_2 at u'_1 and v'_1 respectively and, P_2 intersects C_1 as well as C_2 at u'_2 and v'_2 respectively. If $u_i = u'_i$ and $v_i = v'_i$ for $i = 1, 2$; that is P_i is internally disjoint from C_i for $i = 1, 2$, then there are two $u_1 - u_2$ $(v_1 - v_2)$ paths $C[u_1, u_2]$ $(C[v_1, v_2])$ and $C[u_2, u_1]$ $(C[v_2, v_1])$ in C_1 (C_2). Among these paths, choose, say, $C[u_1, u_2]$ and $C[v_1, v_2]$ with maximum size. Note that,

$$S(C[u_1, u_2]) \geq \frac{1}{2}S(C_1)$$
$$S(C[v_1, v_2]) \geq \frac{1}{2}S(C_2)$$

Now construct a new cycle $C : C[u_1, u_2] \cup P_1 \cup C[v_1, v_2] \cup P_2$. The size of newly constructed cycle is

$$\begin{aligned} S(C) &= S(C[u_1, u_2]) + S(C[v_1, v_2]) + S(P_1) + S(P_2) \\ &> \frac{1}{2}S(C_1) + \frac{1}{2}S(C_2) = S(C_1) \end{aligned}$$

Thus, $S(C) > S(C_i)$, for $i = 1, 2$, a contradiction.

If $u_i \neq u'_i$, $v_i \neq v'_i$ when $i = 1, 2$ then as above, we get two paths say, $C[u'_1, u'_2]$ and $C[v'_1, v'_2]$ from C_1 and C_2 with maximum size. The cycle C' is constructed as $C' : C[u'_1, u'_2] \cup P[u'_1, v'_1] \cup C[v'_1, v'_2] \cup P[u'_2, v'_2]$. Using similar arguments, we can show that $S(C') > S(C_1)$, which is again a contradiction. Hence, C_1 and C_2 have a vertex in common (Fig. 8.18).

Suppose that C_1 and C_2 share only one vertex, say u. Since G has no fuzzy cutvertices, clearly, u cannot be a cut vertex. Thus there is a path, say P having one end u_1 in $C_1 \setminus u$ and the other, say u_2 in $C_2 \setminus u$, disjoint from $C_1 \cup C_2$. Let Q_1 denote the $u_1 - u$ path of C_1 with maximum size, and Q_2 denote the $v_1 - u$ path of C_2 with maximum size. Then $Q_1 \cup Q_2 \cup P$ is a cycle with size strictly grater than size of C_1

(or C_2), which is not possible and therefore, there exist at least two vertices shared by C_1 and C_2. ■

Corollary 8.4.8 *Let G be a 2^*-connected. Then any two heavy fuzzy cycles share at least two vertices.*

8.5 Hamiltonian Fuzzy Graphs

As in classical case, a fuzzy graph is said to be Hamiltonian if it contains a Hamilton cycle [130]. We discuss conditions for a fuzzy graph to have a particular type of Hamilton cycle called as *Hamilton fuzzy cycle*, based on the vertex neighbor sets of the fuzzy graph. We define Hamilton fuzzy cycle as follows.

Definition 8.5.1 Let $G = (\sigma, \mu)$ be a fuzzy graph. A cycle C containing every vertex of σ^* is said to be a *Hamilton fuzzy cycle* if C is a Hamilton cycle in G^* and a fuzzy cycle of G. A Hamilton fuzzy cycle with maximum size in G is called a *heavy Hamilton fuzzy cycle*.

Clearly, every Hamilton fuzzy cycle is a Hamilton cycle. We use the name Hamiltonian to mean that there is a Hamilton fuzzy cycle, unless stated otherwise. Also, Hamiltonian fuzzy graphs are connected and have no cutvertices.

Theorem 8.5.2 *If $|N(v, G)| \geq k > 2$ (k is a positive integer ≥ 2) for every $v \in V(G)$, then G contains a cycle $C = (\tau, \nu)$ with $|\tau^*| \geq k + 1$*

Proof Let $P = (\rho, \xi)$ be the path in G with $|\rho^*|$ maximum in G. Suppose that the vertices are in the order $v_1, v_2, \ldots, v_n$. Then $N(v_1, G) \cup N(v_n, G) \subset \rho^*$. Note that $v_1 \in N(v_i, G)$ for at least k v_i's. Let v_j be the vertex with largest suffix such that $v_1 \in N(v_j, G)$. Then $j \geq k + 1$. Now rearrange the vertices as $v_1, v_2, \ldots, v_j, v_1$ to form a cycle. Let $C = (\tau, \nu)$ be the cycle obtained with $\tau^* = \{v_1, v_2, \ldots, v_j\}$ and $\nu^* = \{\xi(xy) : x, y \in \tau^*\} \cup \mu(v_1 v_j)$ and $|\tau^*| = j \geq k + 1$. ■

Theorem 8.5.3 *If $G = (\sigma, \mu)$ is connected in which $|N(v, G)| \geq k$ (k is a positive integer) for every $v \in \sigma^*$ and $|\sigma^*| \geq 2k$, then G is Hamiltonian.*

Proof Suppose that for $k > 1$, there exists $G' = (\sigma', \mu')$ with $|\sigma'^*| < 2k$ which is not Hamiltonian. Let the cycle $C = (\tau, \nu)$ be contained in G' with $|\tau^*|$ maximum and let the vertices of C be in the order $v_1, v_2, \ldots, v_l$. By Theorem 8.5.2, $|\tau^*| \geq k + 1$ and by hypothesis $|\tau^*| \leq |\sigma'^*| \leq 2k - 1$. At least one of the vertices of τ^* is connected to a vertex of G' which is not in τ^*, and we can take this vertex to be v_l.

Suppose that the path P in G' has more vertices which start from v_l and which have no vertex, except v_l in common with C. Let $P = v_l, v_{l+1}, \ldots, v_{l+m}$. Now we prove that $m \geq k$. Suppose, on the contrary, that $m \leq k - 1$. Since P is chosen in a particular way, v_{l+m} is joined only to vertices among $v_1, v_2, \ldots, v_l v_{l+1}, \ldots, v_{l+m-1}$ and it is therefore joined to at least k of these. Hence v_{l+m} is joined to at least $k - m$ vertices of C besides v_l.

If v_{l+m} is joined to a vertex v_j of C different from v_l, then $j \geq m + 1$. The path whose vertices, in order are $v_{l+m}, v_j, v_{j+1} \ldots, v_l$ and P together form a cycle C' of length $l + m + 1 - j$. Also C is A cycle in G' with maximum $|\tau^*|$, so $l + m + 1 - j < l$ and hence $j > m + 1$. Similarly, $j < l - m - 1$. Thus v_{l+m} is joined to at least $k - m$ of the vertices $v_{m+1}, v_{m+2}, \ldots, v_{l-m-2}, v_{l-m-1}$. These vertices number $l - 2m - 1$.

Now v_{l+m} is not joined to two neighboring vertices of C because if it were C, would clearly not be one of the cycles of G' with maximum $|\tau^*|$. Hence v_{l+m} is joined to at least $l + m$ of the vertices $v_{m+1}, v_{m+2}, \ldots, v_{l-m-2}, v_{l-m-1}$, which number $l - 2m - 1$, but it is not joined to two successive ones. It follows that there are at least $2(k - m) - 1$ of these vertices, so that $l - 2m - 1 \geq 2(k - m) - 1$ and therefore $l \geq 2k$. On the other hand, if $l \geq 2k$, then G is Hamiltonian because $|\sigma^*| \geq 2k$. However, this contradicts the assumption that G is not Hamiltonian.

Thus, $m \geq k$. But G contains at least $l + m$ vertices and $l \geq k + 1$ by Theorem 8.5.2. Hence if $m \geq k$, G contains at least $2k + 1$ vertices, a contradiction. ■

We observe that assumption in Theorem 8.5.3 is only sufficient, but not necessary. This can be seen from the following example.

Example 8.5.4 If $G = (\sigma, \mu)$ is with $\mu(ab) = 0.1$ $\mu(ac) = 0.1$ and $\mu(bc) = 0.7$, then G is a fuzzy cycle and hence is Hamiltonian. For each vertex $v \in \sigma^*$, $|N(v, G)| = 2$ but $|\sigma^*| < 4$.

Theorem 8.5.5 *Let $G = (\sigma, \mu)$ be connected and $|N_s(v, G)| \geq k$ (k is a positive integer) for every $v \in V(G)$ and $|V(G)| \geq 2k$, then G is Hamiltonian.*

The proof of the Theorem 8.5.5 is similar to that of Theorem 8.5.3. Now we can easily observe that a complete fuzzy graph having three or more vertices is Hamiltonian.

Theorem 8.5.6 *Suppose G is a CFG on n vertices and $\sigma(v_1) \leq \sigma(v_2) \leq \cdots \leq \sigma(v_n)$. Then size of a heavy Hamiltonian fuzzy cycle in G is $2\sigma(v_1) + \sigma(v_2) + \cdots + \sigma(v_{n-1})$.*

Lemma 8.5.7 *Suppose that $G = (\sigma, \mu)$ is connected and G^* is 2-connected. Let t be a positive real number. If $d(v, G) \geq t$ for all $v \in \sigma^* \backslash \{x\}$, then G will contain a path of size at least t from x to any other vertex y of G.*

Proof We use induction on $|\sigma^*|$. Let us first consider the case $|\sigma^*| = 3$. Then any path with three vertices will have size more than t. Suppose $|\sigma^*| \geq 4$. Assume that the theorem is true for all fuzzy graphs with $|\sigma^*| - 1$ vertices. Suppose $H = (\sigma', \mu')$ is a fuzzy subgraph. Let $\sigma'^* = \sigma^* \backslash \{x\}$ and $\mu'(e) = \mu(e)$ for every edge e in μ'^*.

Suppose H^* is 2-connected.

Let $t_0 = \max\{\mu(vx) : v \in N(x, G)\}$ and $\mu(x'x) = t_0$ for some $x' \in N(x, G)$. Note that,

$$d(v, H) = d(v, G) \qquad \forall\ v \in \sigma^* \backslash N(x, G) \tag{8.5.1}$$

$$d(v, H) \geq d(v, G) - t_0 \qquad \forall\ v \in N(x, G) \tag{8.5.2}$$

From Eqs. (8.5.1) and (8.5.2), $d(v, H) \geq d(v, G) - t_0 \geq t - t_0\ \forall\ v \in \sigma'^*$ and $d(v, H) \geq t - t_0\ \forall\ v \in \sigma'^* \setminus \{x'\}$. By induction, we can find an $x' - y$ path P' of size at least $t - t_0$ in H. Now set $P = P' \cup \{x'x\}$. Then P is a path of size at least t from x to y.

Suppose H^* is not 2-connected.

Here, H^* may contain cut vertices. Now choose a block $B = (\tau, \nu)$ of H^* with exactly one cutvertex b of H^*. Clearly $x \in N(v, B)$ for some $v \in \tau^*$.

Let $x \in N(v, B) \cap N(b, B)$ for some $v \in \tau^* \setminus \{b\}$ and let $B_1 = (\tau_1, \nu_1)$ induced by $\tau^* \cup \{x\}$. Clearly, B_1^* is 2-connected. Hence, for $v \in \tau_1^* \setminus \{x\}, d(v, B_1) = d(v, G) \geq t$. Again by the induction hypothesis, B_1 contains a path of size at least t from x to a vertex x' of B_1 and hence in G.

If $xb \in \tau^*$ and $x \notin N(v, B)$ for all $v \in \tau^* \setminus \{b\}$,then construct a new fuzzy graph $B_2 = (\tau_2, \nu_2)$ induced by $\tau_2^* = \tau^* \cup \{x\}$ with an additional edge xb_2 whose membership value is $CONN_G(x, b_2)$ for some $b_2 \in N(b, B_2)$. Again B_2^* is 2-connected. For $v \in \tau_2^* \setminus \{x\}$, $d(v, B_2) = d(v, G) \geq t$. Thus, for some $x' \in tau_2^*$, there exists a $x - x'$ path of size t in B_2 and hence in G. ∎

Theorem 8.5.8 *Suppose G is a connected fuzzy graph with G^* 2-connected. Let t be a positive real number. If $d(v, G) \geq t$ for every $v \in V(G)$, then G has either a cycle of size at least 2t or a Hamilton fuzzy cycle.*

Proof Suppose $C : (\tau, \nu)$ is a heavy cycle in G which is not Hamilton. Consider a component $H = (\sigma_1, \mu_1)$ of $G - C$. Then H may be separable or non-separable. Let P be an $x - y$ path in H. By Lemma 8.5.7, we can choose P in such a way that $S(P) \geq t - d(y, G) \geq t - d(y, C)$ (Fig. 8.19).

Suppose H is not a block.

Suppose $B_1 : (\tau_1, \nu_1)$ and $B_2 : (\tau_2, \nu_2)$ are two different endblocks of H and that b_i is the unique cut vertex in B_i for $i = 1, 2$. Now choose $x_i \in \tau_i^* \setminus \{b_i\}$ for $i = 1, 2$ in such a way that

1. $d(x_i, C) > 0$ and

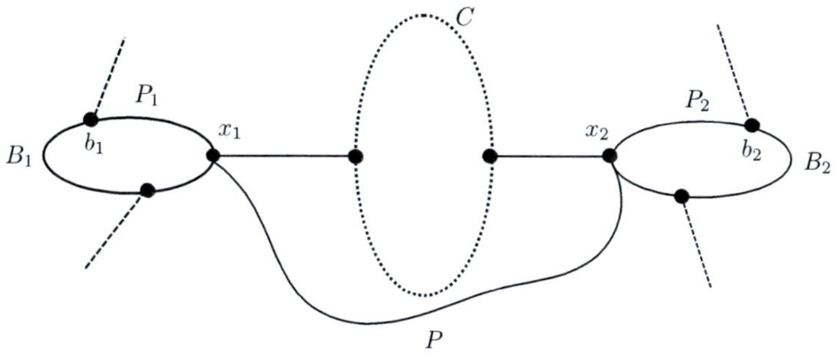

Fig. 8.19 Construction of path P in Case 1

2. $d(x_i, C) \geq d(v, C)$ for all $v \in \sigma_1^* \setminus \{x, y\}$.

Then clearly,

$$\begin{aligned} d(v, B_i) &= d(v) - d(v, C) \\ &\geq t - d(x_i, C) \qquad \forall\, v \in \tau_i^* \setminus \{b_i\} \end{aligned}$$

For $i = 1, 2$, we apply Lemma 8.5.7 to B_i and get a (x_i, b_i)- path say P_i, in B_i of size,

$$S(P_i) \geq t - d(x_i, C) \tag{8.5.3}$$

If $|N(x, C)| \cup |N(y, C)| \geq 2$, then we let P be an (x_1, x_2)- path of maximum size in H. Then, by (8.5.3)

$$S(P) \geq S(P_1) + S(P_2) \geq t - \min\{d(x_1, C), d(x_2, C)\}.$$

By Corollary 8.4.6,

$$S(C) \geq 2\min\{d(x_1, C), d(x_2, C)\} + 2S(P) \geq 2t.$$

If $N^C(x_1) = N^C(x_2) = \{u_1\}$, then there exists a vertex $u_2 \in \tau^* \setminus \{u_1\}$ adjacent to some vertex $b \in \sigma_1^*$. Since $(\tau_1^* \setminus \{b_1\}) \cap (\tau_2^* \setminus \{b_2\}) = \emptyset$, b cannot belong to both $(\tau_1^* \setminus \{b_1\})$ and $(\tau_2^* \setminus \{b_2\})$. Suppose that $b \notin (\tau_2^* \setminus \{b_2\})$. Extending the path P_2 in H from b_2 to b, we obtain an (x_2, b)-path $P^{'}$ in H of size

$$S(P') \geq S(P_2). \tag{8.5.4}$$

Since C is heavy,

$$\begin{aligned}S(C) &= S(C[u_1, u_2]) + S(C[u_2, u_1]) \\ &\geq 2(\mu(u_1x_2)) + S(P') + \mu(u_2b) \\ &\geq 2(\mu(u_1x_2)) + S(P')\end{aligned}$$

Now using (8.5.3) and (8.5.4), for $i = 2$, we have

$$S(C) = 2(\mu(u_1x_2) + S(P_2)) \geq 2t + 2(\mu(u_1x_2) - d(x_2, C)) = 2t.$$

Suppose H is a block.

If $|\sigma_1^*| = 1$, then $S(P) = 0 \geq t - d_s(y, G) \geq t - d_s(y, C) \geq 2t$.

Let $\sigma_1^* = x$ and $N(x, C) = \{v_1, v_2, \ldots, v_n\}$ $(n \geq 2)$, where $v_i \in \tau^*$ for $i = 1, 2, \ldots, n$. For a pair (v_i, v_{i+1}), we form a cycle C_i using the cycle C by completely replacing the segment $C[v_i, v_{i+1}]$ with path $v_i - x - v_{i+1}$. There exist n distinct pairs (v_i, v_{i+1}) of vertices for $i = 1, 2, \ldots, n$. Thus we get n distinct cycles, with $S(C_i) \leq S(C)$. Together we have,

$$\sum i = 1nS(C_i) = (n-1)S(C) + 2d(x, C) \tag{8.5.5}$$

$$\sum i = 1nS(C_i) \leq nS(C) \tag{8.5.6}$$

From Eqs. (8.5.5) and (8.5.6), we can conclude that, $S(C) \geq 2d(x, C)$, which gives $S(C) \geq 2t$.

If suppose $|\sigma_1^*| > 1$, then find a path say, P from x to y in such a way that

1. $|N(x, C)| \geq 1, |N(y, C)| \geq 1$ and
2. $d(x, C) \geq d(y, C) \geq d(v, C)$ for all $v \in \sigma_1^* \setminus \{x, y\}$.

If $|N(x, C)| \cup |N(y, C)| \geq 2$, then, by Corollary 8.4.6,

$$\begin{aligned}S(C) &\geq 2\min\{d(x, C), d(y, C)\} + 2S(P) \\ &\geq 2d(y, C) + 2S(P) \\ &\geq 2t\end{aligned}$$

Now assume $N(x, C) \cup N(y, C) = \{p\}$. Since G^* is 2-connected, there exists a vertex $q \in \tau^* \setminus \{p\}$, adjacent with vertex $z \in \sigma_1^* \setminus \{x, y\}$ (Fig. 8.20).

By the selection of vertices x and y,

$$d(v, H) = d(v, G) - d(v, C) \geq t - d(v, C) \quad \forall\ v \in \sigma_1^*.$$

By applying Corollary 8.4.6 to the fuzzy subgraph H , we have an (x, z)-path P' in H of size

$$S(P') \geq t - d(v, C) = t - \mu(xp).$$

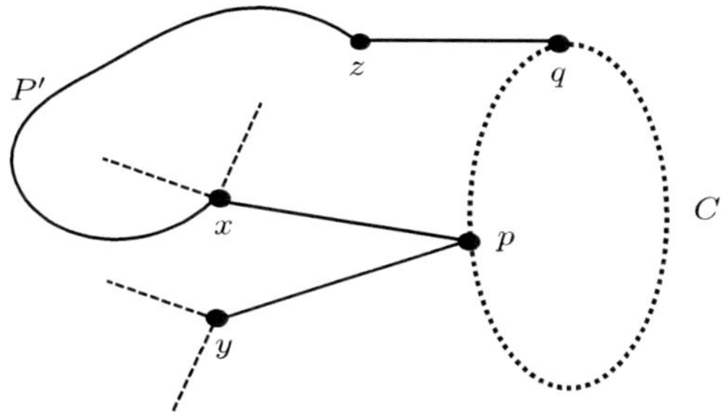

Fig. 8.20 Construction of path P' in Case 2

Hence,

$$S(C) = S(C[p, q]) + S(C[q, p]) \geq 2(\mu(px) + S(P') + \mu(qz)) \geq 2t.$$

■

8.6 Algorithms for Hamiltonian Fuzzy Graphs

We now discuss algorithms for finding heavy paths, heavy cycles and Hamiltonian cycles in a fuzzy graph.

Algorithm 8.6.1 Algorithm to find heavy path with a given initial vertex u.

Let $G = (\sigma, \mu)$ have support $\sigma^* = \{v_1, v_2, \ldots, v_n\}$ and adjacency matrix $A(G)$. Let $u = v_i$.

Step 1. Construct $A(G)$.

Step 2. Look for the first element, say a_{ij} with $a_{ij} > 0$, in the ith row of $A(G)$ (If all entries are zero, then conclude that there is no path with initial vertex v_i).

Step 3. Identify the jth row corresponding to v_j. If all entries except a_{ji} are zeros, then the required path is $v_i - v_j$, else, this path can be extended longer.

Step 4. Let a_{jk} be the selected non- zero occurrence in the jth row.

Step 5. Consider kth row. Determine whether the path $v_i - v_j - v_k$ can be extended or not. If not, go to Step 3 and find an element greater than 0, except a_{ji} and a_{jk} and continue the iteration to find new paths.

Step 6. Else, repeat Step 3, 4 and 5 until we get a zero entry or forms a path with repetition of vertices. If path P_1 is obtained, then find $S(P_1)$. Go to step 5 and identify the next non-zero element in the ith row other than a_{ij}.

Step 7. Repeat Step 3–5 to find all other paths with initial vertex as v_i.

Step 8. Let $P_1, P_2, \ldots, P_l$ be the paths with initial vertex as v_i. Find $S(P_i)$ for $i = 1, 2, \ldots, l$.

Step 9. Identify the path P_r with $S(P_r) = \max\{S(P_i)\}_1^l$, if exists.

Algorithm 8.6.1 is explained in Example 8.6.2.

Example 8.6.2 Suppose G has a support on 4 vertices, v, w, x, y. Let $\mu(wv) = 0.1$, $\mu(wy) = 0.45$, $\mu(vy) = 1$ and $\mu(yx) = 0.3$. The adjacency matrix of G is given below.

$$A(G) = \begin{array}{c} \\ u \\ w \\ v \\ y \\ x \end{array} \begin{array}{c} \begin{array}{ccccc} u & w & v & y & x \end{array} \\ \begin{bmatrix} 0 & 0.09 & 0 & 0 & 0 \\ 0.09 & 0 & 0.1 & 0.45 & 0 \\ 0 & 0.1 & 0 & 1 & 0 \\ 0 & 0.45 & 1 & 0 & 0.3 \\ 0 & 0 & 0 & 0.3 & 0 \end{bmatrix} \end{array}$$

Start with the adjacency matrix $A(G)$. Find a heavy path P^u, with u as the initial vertex. The first entry, non-zero, appearing in the uth row is 0.09 and this corresponds to the row u and column w, that is, from u it reaches w, hence the path is $u - w$.

In row w, the entries which are non-zero are 0.09, 0.1 and 0.45, appears in the columns u, v and x respectively. We move with v. So we have the path $u - w - v$ (u is not selected since $u - w - u$ is will not form a path).

In the row v, the non-zero entries are 0.1 and 0.45 representing columns w and y. Selection of w is not possible so we move with y. So we have the path $u - w - v - y$.

Again in row y, the non-zero entries are 0.45, 1 and 0.3 representing columns w, v and x. Selection of w and v are not possible so we move with x. So we have the path $u - w - v - y - x$.

Now in row x, there is only one non zero entry, namely 0.3, which represents y. Path $u - w - v - y - x$ cannot extend here after. So let P_1^u be the path $u - w - v - y - x$. Size of P_1^u is 1.49.

Now, consider the other non-zero entry 0.45 in the row w, which represents the column y. Preceding with y as above we obtain paths $P_2^u : u - w - y - v$ and $P_3^u : u - w - y - x$ with sizes 1.54 and 0.84 respectively. Hence heavy path in G with u as the initial vertex is $P_2^u : u - w - y - v$.

Repeat Steps 1 through 5 to obtain paths $P_1^w : w - y - x$, $P_2^w : w - u$ and $P_3^w : w - v - y - x$ with w as the initial vertex. Heavy path with w as initial vertex is $P_3^w : w - v - y - x$ whose size is 1.4. Similarly, $v - y - x$, $y - v - w - u$ and $x - y - v - w - u$ are heavy paths starting with v, y and x respectively.

Algorithm 8.6.3 Algorithm to find heavy paths in a fuzzy graph.

Suppose that $G = (\sigma, \mu)$ is connected with support $\sigma^* = \{v_1, v_2, \ldots, v_n\}$.

Step 1. For each vertex $v_i \in \sigma^*$, identify heavy paths P_r with v_r as the initial vertex, for $r = 1, 2, \ldots, n$ using Algorithm 5.1.
Step 2. Find size of each paths P_r.
Step 3. Identify a path P with size $\max\{S(P_r) : r = 1, 2, \ldots, n\}$.
Step 4. Then P is a heavy path.

Illustration of Algorithm 8.6.3. Consider Example 8.6.2. G is a fuzzy graph with 5 vertices. Using Algorithm 5.1, we obtain heavy paths $P_1 : u - w - y - v$, $P_2 : w - v - y - x$, $P_3 : v - y - x$, $P_4 : y - v - w - u$ and $P_5 : x - y - v - w - u$ with initial vertices u, w, v, y and x respectively. Next step is to find sizes of these five paths. $S(P_1) = 1.54$, $S(P_2) = 1.4$, $S(P_3) = 1.3$, $S(P_4) = 1.19$ and $S(P_5) = 1.49$.

$\max\{S(P_r) : r = 1, 2, \ldots, 5\} = 1.54$. Path that gives the size 1.54 is $u - w - y - v$. Therefore, $P = u - w - y - v$ is a heavy path.

In order to proceed to the next algorithm, we need another definition.

Suppose $G = (\sigma, \mu)$ is an f-graph with support $\{v_1, v_2, \ldots, v_n\}$. The *strong adjacency matrix*, $A_s(G)$ is defined as.

$$a_{ij} = \begin{cases} \mu(v_i v_j) & \text{if } v_i v_j \in \mu^* \text{ and } CONN_G(v_i, v_j) \geq \mu(v_i v_j) \\ 0 & \text{otherwise,} \end{cases}$$

$$A_s(G) = \begin{array}{c} \\ v_1 \\ v_2 \\ v_3 \\ v_4 \end{array} \begin{array}{c} v_1 \; v_2 \; \ldots \; v_n \\ \begin{bmatrix} 0 & a_{12} & \ldots & a_{1n} \\ a_{21} & 0 & \ldots & a_{2n} \\ \vdots & \vdots & \ldots & \vdots \\ a_{n1} & a_{n2} & \ldots & 0 \end{bmatrix} \end{array}$$

Algorithm 8.6.4 Algorithm to find a spanning path that gives a Hamiltonian fuzzy cycle in G.

Let G be with $|\sigma^*| \geq 3$ and $A_s(G)$ its strong adjacency matrix. Find a minimum element, which is non-zero from the strong adjacency matrix. Let it be $a_{ij} = \mu(v_i v_j)$. This gives a path starting at v_i and ending at v_j. Now from the row corresponding to v_j find a suitable non-zero element. This process is repeated till the resultant path is a spanning path which can be enlarged to a Hamiltonian fuzzy cycle.

Step 1. Find $A_s(G)$.

Step 2. Find an element with the smallest value (> 0) in $A_s(G)$, say $a_{ij} = x_1$. (If more than one is available, then consider any one of them).

Step 3. If the selected x_1 does not allow for a path with no replication of vertices, then choose the very next element with higher value.

Step 4. Recognize the row and column represented by x_1, say v_i and v_j.

Step 5. Repeat Step 2 in the row of v_j, forming a path (an appropriate element from the row is picked to make a $v_i - v_j$ path).

Step 6. Return to Step 3 and follow Step 4 row by row till a spanning path is discovered. Else, conclude that there are no spanning paths and redo Step 2 or Step 3 as needed. If a spanning path P is obtained, then one row will not contain any marked entries.

Step 7. Identify and choose an entry (>0) from this row, if exists, to form a Hamiltonian fuzzy cycle.

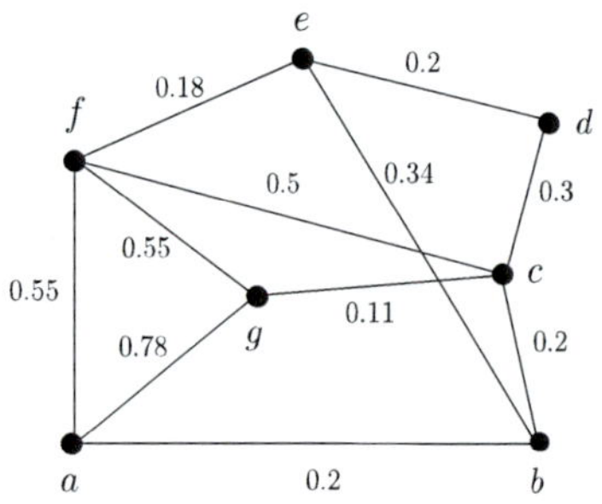

Fig. 8.21 Fuzzy graph G in illustration of Algorithm 8.6.4

Illustration of Algorithm 8.6.4. Consider G in Fig. 8.21 with $\mu(ab) = \mu(ed) = \mu(c) = 0.2,\ \mu(af) = \mu(fg) = 0.55,\ \mu(ef) = 0.18,\ \mu(cf) = 0.5,\ \mu(ce) = 0.11,$ $\mu(eb) = 0.34\ \mu(cd) = 0.3$ and $\mu(ag) = 0.78$. The heavy path P_h in G is $a - c - e - d - f$ with size 2.3 and $a - c - e - a$ is a heavy cycle with size 2.25. Heavy fuzzy cycles in G are $a - b - c - d - f - e - a$ and $e - c - d - f - e$ with sizes 1.4 each.

The strong adjacency matrix $A_s(G)$ of G is given by,

$$A_s(G) = \begin{array}{c} \\ a \\ b \\ c \\ d \\ e \\ f \\ g \end{array} \begin{array}{c} \begin{array}{ccccccc} a & b & c & d & e & f & g \end{array} \\ \begin{bmatrix} 0 & 0.2 & 0 & 0 & 0 & 0.55 & 0.78 \\ 0.2 & 0 & 0.2 & 0 & 0.34 & 0 & 0 \\ 0 & 0.2 & 0 & 0.3 & 0 & 0.5 & 0 \\ 0 & 0 & 0.3 & 0 & 0.2 & 0 & 0 \\ 0 & 0.34 & 0 & 0.2 & 0 & 0 & 0 \\ 0.55 & 0 & 0.5 & 0 & 0 & 0 & 0.55 \\ 0.78 & 0 & 0.3 & 0 & 0 & 0.55 & 0 \end{bmatrix} \end{array}$$

Form the matrix $A_s(G)$. Observe that the non-zero minimum of the first row appears in the second column. Choose the entry as, $a_{12} = 0.2$. From a, we reach b through the edge ab. In b, the minimum non-zero element 0.02 appears in the columns a and c. We go with c. So we have the path $(a - b - c)$. Repeating Step 3 and Step 4, we obtain the path $a - b - c - d - e - a$, which is not a spanning path. Repeating the process with another non zero minimum of of row b, that is with, $b_{25} = 0.3$, represented by the row e and column d, we obtain the path $a - b - d$. Now repeating Steps 3 and 4, we get a path $a - b - e - d - c - f - g - a$, which is a Hamiltonian fuzzy cycle with size 2.09.

Algorithm 8.6.5 Algorithm for finding Hamiltonian fuzzy cycles using strong degree of vertices.

This algorithm is used to find fuzzy cycles, which are Hamilton, using strong degrees $d_s(v, G)$ of vertices in G. Select one vertex possessing minimum strong degree and follow the algorithm.

Step 1. For each $v \in \sigma^*$, find the strong degree $d_s(v, G)$.

Step 2. Choose one vertex having strong degree $\delta_s(G)$ as the initial vertex of our path. (If there exist more than one vertex with strong degree equal to $\delta_s(G)$, then begin with any of them).

Step 3. Identify the vertex that belongs to the strong neighbor set of the chosen vertex whose strong degree is next higher to the strong degree of the vertex selected in Step 2.

Step 4. Extend the path through the new vertex obtained.

Step 5. Approach the vertex that is unvisited, adjacent to, and strong degree greater than that of the vertex obtained in Step 3. (If there exist more than one vertex with same characteristics, then extend the path with any of them).

Step 6. Repeat Step 3 followed by Step 4 and Step 5 till a Hamiltonian fuzzy cycle is found. Else, repeat Step 2 or Step 3 accordingly.

The above algorithm is illustrated in the following example.

Consider the fuzzy graph in Fig. 8.21, the strong degree of the vertices are $d_s(a) = 0.98$, $d_s(b) = 0.74$, $d_s(c) = 1$, $d_s(d) = 0.5$, $d_s(e) = 0.54$, $d_s(f) = 1.6$ and $d_s(g) = 1.33$. Select the vertex d, as it has the minimum strong degree $d_s(d) = 0.5$. The unvisited vertices of the strong neighbor set of d are e and c with $d_s(e) = 0.54$ and $d_s(c) = 1$. Minimum of these values is 0.54 which represents the vertex e. Therefore, the path travels from d to e that is $d - e$. Repeat Steps 3 and 4. The only strong neighbor of e is b, hence the new path is $d - e - b$. Again repeat Steps 3 and 4. The unvisited vertices of strong neighbor set of b are a and c with $d_s(a) = 0.98$ and $d_s(c) = 1$. Minimum among these values is 0.98 which represents the vertex a. So the path extends to $d - e - b - a$. Repeating this process we obtain a spanning path $d - e - b - a - g - f - c$. Note that $cd \in \mu^*$, hence this spanning path can be extended to spanning cycle. Thus the cycle $d - e - b - a - g - f - c - d$ is the unique Hamiltonian fuzzy cycle and hence G is Hamiltonian.

8.7 A Model for Human Trafficking

We use a model provided by Mordeson and Mathew [109] for the determination of illegal immigration routes related to human trafficking. The model is in the form of a *directed fuzzy graph(DFG)*. The concepts discussed in this paper can be extended to directed fuzzy graphs. A *directed path (di-path)* is called a *heavy di-path* if it is a directed path of maximum size. A directed path is called a *Directed Hamiltonian path* if it passes through each of the vertices of the DFG exactly once, where as a *Hamiltonian fuzzy cycle* is a directed fuzzy cycle that visits every vertex in the fuzzy graph once without repetitions. A Hamiltonian cycle becomes Hamilton fuzzy cycle if it is also a directed fuzzy cycle.

The data of illegal human flow between different parts of the globe, given in Table 8.2 is taken from [109]. We consider the flow graph $G = (\sigma, \mu)$ with the vertex set $\sigma^* = \{a, b, c, d, e, f, g, h, i, j\}$, obtained in [109].

Table 8.2 Illegal emigration between 10 different locations

Region	a	b	c	d	e	f	g	h	i	j
West & Cen. Europe(a)	0.62	0.13								
W. South Europe(b)		0.16								
Cen. Europe & Bal.(c)		0.27	0.79		0.05					
East. Eur. & Cen. Asia(d)	0.04	0.04	0.05	0.99						0.06
N & C Amer. and Car.(e)	0.08	✓	✓		0.59	0.04				
South America(f)		0.07			0.03	0.94				
East Asia & Pac.(g)	0.07	0.07	✓		0.25	0.01	0.97			0.33
South Asia(h)		✓			0.07			0.96		0.18
Sub Saharan Africa(i)		0.16						1.0		0.10
Middle East(j)										0.31

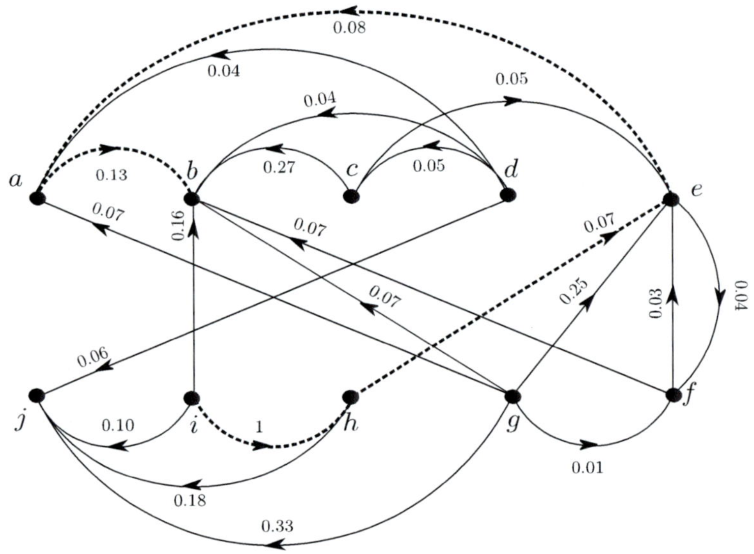

Fig. 8.22 Fuzzy graph representation of human trafficking data

The normalized data flow of illegal immigrants can be studied using the directed fuzzy graph model G as in Fig. 8.22.

If G is a DFN with $\sigma^* = \{v_1, v_2, \ldots, v_n\}$, then the directed adjacency matrix, denoted by $A(\overrightarrow{G})$ is an $n \times n$ matrix where,

$$a_{ij} = \begin{cases} \mu(v_i v_j) & \text{if } i \neq j \text{ and } (v_i v_j) \in \mu^* \\ 0 & \text{otherwise.} \end{cases}$$

$$A(\overrightarrow{G}) = \begin{array}{c} \\ a \\ b \\ c \\ d \\ e \\ f \\ g \\ h \\ i \\ j \end{array} \begin{array}{c} \begin{array}{cccccccccc} a & b & c & d & e & f & g & h & i & j \end{array} \\ \begin{bmatrix} 0 & 0.13 & 0 & 0 & 0 & 0 & 0 & 0 & 0 & 0 \\ 0 & 0 & 0 & 0 & 0 & 0 & 0 & 0 & 0 & 0 \\ 0 & 0.27 & 0 & 0 & 0.05 & 0 & 0 & 0 & 0 & 0 \\ 0.04 & 0.04 & 0.05 & 0 & 0 & 0 & 0 & 0 & 0 & 0.06 \\ 0.08 & 0 & 0 & 0 & 0 & 0.04 & 0 & 0 & 0 & 0 \\ 0 & 0.07 & 0 & 0 & 0.03 & 0 & 0 & 0 & 0 & 0 \\ 0.07 & 0.07 & 0 & 0 & 0.25 & 0.01 & 0 & 0 & 0 & 0.33 \\ 0 & 0 & 0 & 0 & 0.07 & 0 & 0 & 0 & 0 & 0.18 \\ 0 & 0.16 & 0 & 0 & & 0 & 0 & 1 & 0 & 0.10 \\ 0 & 0 & 0 & 0 & 0 & 0 & 0 & 0 & 0 & 0 \end{bmatrix} \end{array}$$

For each $v \in \sigma^*$, use Algorithm 5.1 to find heavy paths with v as the initial vertex. With the help of Algorithm 5.2 we obtain $i \to h \to e \to a \to b$ as the only heavy path in G with size 1.28. From this DFN, it follows that maximum human trafficking route is starting from the Sub Saharan Africa going through South Asia, North and Central America and the Caribbeans to the West and Central Europe. Note that the directed fuzzy graph in Fig. 8.22 is a directed fuzzy tree and hence G is not a Hamiltonian fuzzy graph.

Chapter 9
Connectivity and Human Trafficking

The concept of connectivity has been studied deeply in fuzzy graph theory. Several connectivity parameters have been defined, studied and applied. Most of the wide dynamic networks like internet, follow a fuzzy graph model, in a strict sense. Both introductory papers by Rosenfeld and Yeh and Bang concentrated at this theme. This chapter is devoted to some connectivity parameters and their applications.

9.1 Fuzzy Vertex and Edge Connectivity Parameters

This section contains some results on fuzzy vertex connectivity and fuzzy edge connectivity, defined in [98]. Fuzzy edge connectivity of saturated fuzzy cycles and complements of cycles are some of the topics discussed in this section. Section 8.1 presents generalized connectivity parameters and finds their relationships with the existing parameters. Section 8.2 gives an algorithm for fuzzy graph clustering using generalized connectivity parameters and Sect. 8.3 contains an application related to human trafficking. The results in these sections depend on [151].

In [98], it is shown that if $G = (\sigma, \mu)$ is a fuzzy graph and $H = (\sigma, \nu)$ is a partial fuzzy subgraph of G, then $\kappa(H) \leq \kappa(G)$. This result is not generally true for edge connectivity. But is true in some situations as seen in the following result.

Theorem 9.1.1 *Let $G = (\sigma, \mu)$ be a fuzzy graph with $|\sigma^*| = n$. If $H = (\sigma, \nu)$ is a partial fuzzy subgraph having the same vertex set of G, then $\kappa'(H) \leq \kappa'(G)$.*

The vertex set of H is the same as the vertex set of G in Theorem 9.1.1. This result is not true for every partial fuzzy subgraph of G. Consider the fuzzy graph shown in Fig. 9.1.

J. N. Mordeson and S. Mathew, *Mathematics of Uncertainty for Coping with World Challenges*, Studies in Systems, Decision and Control 353,
https://doi.org/10.1007/978-3-030-68684-0_9

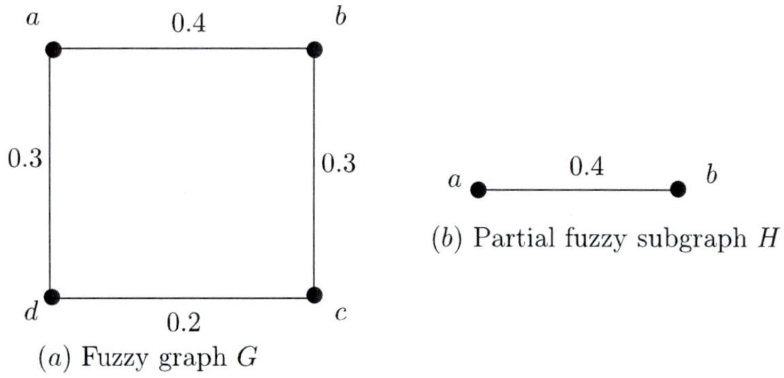

Fig. 9.1 Partial fuzzy subgraph H with $\kappa'(H) > \kappa'(G)$

In Fig. 9.1, G is a fuzzy graph with $\kappa'(G) = 0.3$. H is a partial fuzzy subgraph of G with $\kappa'(H) = 0.4$. Also we can see that removal of an edge never enhances the edge connectivity of a fuzzy graph as seen from Theorem 9.1.2.

Theorem 9.1.2 *Let $G = (\sigma, \mu)$ be a fuzzy graph. Then $\kappa'(G - e) \leq \kappa'(G)$ for every $e \in \mu^*$.*

It is not hard to find a proof for Theorem 9.1.2 using Theorem 9.1.1 and the fact that $G - e$ is a partial fuzzy subgraph of G on the same vertex set. Now we construct a super fuzzy graph whose edge connectivity is more than that of a given fuzzy graph.

Theorem 9.1.3 *For any connected fuzzy graph with fuzzy edge connectivity $\kappa'(G)$, there exists a connected super fuzzy graph G' of G with fuzzy edge connectivity $\kappa'(G') > \kappa'(G)$.*

Proof Suppose that $\sigma^* = \{u\}$, $\kappa'(G) = 0$. Construct a super fuzzy graph as follows: Let $G' = (\rho, \nu)$ as $\rho^* = \sigma^* \cup \{v\}$ and $\nu^* = \{uv\}$. Define,

$$\rho(x) = \begin{cases} \sigma(x) & if x = u \\ 1 & x = v \\ \end{cases}$$

and $\nu(uv) = t,\ t \in (0, 1]$. Clearly $\kappa'(G') > \kappa'(G)$.

Let G be fuzzy graph with $\sigma^* = \{u, v\}$, and $\mu^* = \{uv\}$. Then $\kappa'(G) = \mu(uv)$. Let $\mu(uv) = t$. Construct a super graph $G' = (\rho, \nu)$ with $\rho^* = \sigma^* \cup \{z\}$ and $\nu^* = \mu^* \cup \{uz, vz\}$, where

$$\rho(x) = \begin{cases} \sigma(x) & if\ x \in \sigma^* \\ 1 & if\ x = z \\ \end{cases}$$

and

$$\nu(xy) = \begin{cases} \mu(xy) & if\ xy = uv \\ s & if\ xy = uz \text{ or } xy = vz,\ s \nleq t. \end{cases}$$

Suppose that $s > t$, Then, $\kappa'(G') = s > t = \kappa'(G)$, else $\kappa'(G') = 2s > t = \kappa'(G)$.

Suppose that G is a complete fuzzy graph with n vertices say $\{v_1, v_2, \ldots, v_n\}$, $n \geq 3$. Then $\kappa'(G) = n * \wedge\{\sigma(x) : x \in \sigma^*\}$. Construct a super graph $G' = (\rho, \nu)$ such that $\rho^* = \sigma^* \cup \{z\}$ and $\nu^* = \mu^* \cup \{v_i z : i = 1, 2, \ldots n\}$, where

$$\rho(x) = \begin{cases} \sigma(x) & if\ x \in \sigma^* \\ \wedge\{\sigma(x) : x \in \sigma^*\} & if\ x = z \end{cases}$$

and

$$\nu(xy) = \begin{cases} \mu(xy) & if\ xy \in \mu^* \\ \sigma(z) & otherwise. \end{cases}$$

Then G' is a CFG with $n + 1$ vertices. Hence, $\kappa'(G') = (n + 1) * \wedge\{\sigma(x) : x \in \rho^*\} > n * \wedge\{\sigma(x) : x \in \sigma^*\} = \kappa'(G)$.

Suppose that G is not a complete fuzzy graph with edge connectivity $\kappa'(G)$. Let E be a minimum fuzzy edge cut of G and X be the collection of end vertices of E. Then there exists a pair of vertices $u,\ v$ such that $CONN_{G-E}(u, v) < CONN_G(u, v)$, with at least one of u or v different from the end vertices of E. If uv is a $\delta-$edge, then there exist k internally disjoint strongest $u - v$ paths and uv is the weakest edge of a $k-$cycle. Thus any minimum $u - v$ separating set has cardinality k. Construct a super fuzzy graph $G' = (\rho, \nu)$ with $\rho^* = \sigma^* \cup \{z\}$ and $\nu^* = \mu^* \cup \{uz, vz\}$ such that

$$\rho(x) = \begin{cases} \sigma(x) & if\ x \in \sigma^* \\ 1 & if\ x = z \end{cases}$$

and

$$\nu(xy) = \begin{cases} \mu(xy) & if\ xy \in \mu^* \\ CONN_G(u, v) & if\ x \notin X\ or\ y \notin X. \end{cases}$$

By construction, $u - v$ has $k + 1$ internally disjoint strongest paths. Hence, minimum $u - v$ separating set E' has exactly $k + 1$ edges. Moreover, $s'(E') > s'(E)$. If $\kappa'(G') > \kappa'(G)$, then G' becomes the super graph. However, if $\kappa'(G') = \kappa'(G)$, then let E'' be a minimum fuzzy edge cut of G'. Then there exists a pair of vertices $u_1,\ v_1$ in G' such that $CONN_{G-E}(u_1, v_1) < CONN_G(u_1, v_1)$. If $u_1 v_1$ is a $\delta-$edge or u_1 and v_1 are not adjacent, then the above procedure can be repeated. Suppose G' is disconnected. Let Y be a minimum separating set in G' such that $G' - E$ is disconnected.

Let $G'' = (\rho', \nu')$ be such that $\rho'^* = \rho^* \cup \{z\}$ and $\nu'^* = \nu^* \cup \{u_2z, v_2z\}$, where u_2 and v_2 are in different components with minimum strength of connectedness and

$$\rho'(x) = \begin{cases} \sigma(x) \ if \ x \in \rho^* \\ 1 \quad\quad if \ x = z \end{cases}$$

and

$$\nu'(xy) = \begin{cases} \mu(xy) & if \ xy \in \nu^* \\ CONN_G(u_2, v_2) & if \ xy = u_2z \ or \ xy = v_2z. \end{cases}$$

Let Y' be a minimum $u_2 - v_2$ strength reducing set and there exist $Y + 1$ internally disjoint strongest $u_2 - v_2$ paths in G''. Hence, $s'(Y') > s'(Y)$. If $\kappa'(G'') > \kappa'(G')$ then choose super graph as G''. Suppose $\kappa'(G'') = \kappa'(G')$. Repeat the above procedure that have been already discussed. ■

Theorem 9.1.4 considers the existence of $t-$edge connected complete fuzzy graphs.

Theorem 9.1.4 *There exists a $t-$edge connected complete fuzzy graph for every $t \in \mathbb{R}^+$.*

Proof Assume that $t \in (0, 1]$. A $t-$connected complete fuzzy graph is simply $G = K_2$ with $\sigma^* = \{u, v\}$ and $\mu(uv) \geq t$.

Suppose $t \geq 1$. Let $t \in [n, n + 1)$ for some $n \in \mathbb{N}$. Construct a fuzzy graph G with at least $n + 2$ vertices namely $v_1, v_2, \ldots, v_{n+2}, \ldots, v_{n+k}$ as follows. Let $\sigma(v_i) \geq \frac{t}{(n + k - 1)}$ for $i = 1, 2, \ldots, n + k$, and $\mu(v_iv_j) = \sigma(v_i) \wedge \sigma(v_j)$ for $i \neq j$. Then G is a CFG with $n + k$ vertices. We can assume, without loss of generality, that v_1 is the vertex of minimum degree. Then by Theorem 4 in Ref. [98], $\kappa'(G) = (n + k - 1) * \wedge\{\sigma(x) : x \in \sigma^*\}$. Hence, $\kappa'(G) \geq (n + k) * \frac{t}{(n + k - 1)} = t$. Thus, G is a $t-$edge connected complete fuzzy graph. ■

The next result provides a method to construct super fuzzy graphs for a given fuzzy graph without changing the connectivity values.

Theorem 9.1.5 *Let G be a $t-$connected fuzzy graph and let E be a minimum fuzzy edge cut of G. Let G' be a fuzzy graph obtained by adding a new vertex z to G and joining z to the end vertices of E. Let $\mu(v_iz) = \wedge\{\mu(v_iu) : v_iu$ is a strong edge in $G\}$. Then G' is $t-$edge connected.*

Proof Let $G = (\sigma, \mu)$ be a $t-$edge connected fuzzy graph and E be a fuzzy edge cut with minimum strong weight and X be the set of end vertices of edges in E. Let G' be the fuzzy graph obtained by adding a new vertex z to $G = (\rho, \nu)$ and joining

z to X and $\mu(v_i z) = \wedge\{\mu(v_i u) : v_i u$ is a strong edge in $G\}$. Let E' be the minimum fuzzy edge cut of G'.

Clearly, the edges of $\nu^* \setminus \mu^*$ are all strong. Now suppose E' contains edges from $\nu^* \setminus \mu^*$. Then removal of those edges from E' is a fuzzy edge cut for G. Suppose E' contains no edges from $\nu^* \setminus \mu^*$. Then E' itself is a fuzzy edge cut of G. Thus, $s'(E') > \kappa'(G)$. In both cases, $s'(E') > \kappa'(G)$, Thus, $\kappa'(G') \geq \kappa'(G)$. Hence, G' is $t-$edge connected. ■

Theorem 9.1.6 determines the fuzzy edge connectivity of $\beta-$saturated fuzzy cycles.

Theorem 9.1.6 *Let $G = (\sigma, \mu)$ be a fuzzy cycle with $|\sigma^*| \geq 3$. If G is $\beta-$saturated, then $\kappa'(G) = 2\mu(xy)$, where xy is a weakest edge of G.*

Proof Let $G = (\sigma, \mu)$ be a $\beta-$saturated fuzzy cycle with $|\sigma^*| \geq 3$. Then every vertex is incident with at least one $\beta-$strong edge. The removal of an $\alpha-$strong edge (if exists) affects only the connectivity of its end vertices, where as the removal of a $\beta-$strong edge does not affect the connectivity in G. Thus the removal of a single edge does not affect the edge connectivity between any pair of vertices in G. Also, the removal of any two edges from G disconnects the graph. Thus any two edges form a fuzzy cut set of G, of which one having any two $\beta-$strong edges will have the minimum size. Hence, $\kappa'(G) = 2\mu(xy)$. ■

We next determine the fuzzy edge connectivity of odd cycles which are $\alpha-$saturated.

Theorem 9.1.7 *Let $G = (\sigma, \mu)$ be an odd fuzzy cycle with $|\sigma^*| \geq 3$. If G is $\alpha-$saturated and e is a weakest edge of G, then $\kappa'(G) = \wedge\{m, 2\mu(e)\}$, where $m = \wedge\{\mu(uv) :$ either u or v is incident with $\alpha-$strong edges only$\}$.*

Proof Let $G = (\sigma, \mu)$ be an odd fuzzy cycle with $|\sigma^*| \geq 3$. Let X be the collection of vertices in G which are incident with $\alpha-$strong edges and no $\beta-$strong edges. Since G is $\alpha-$saturated and is an odd cycle, $|X| \neq 0$. Let $u \in X$ and $v \in \sigma^*$ be such that $uv \in \mu^*$. Note that uv is an $\alpha-$strong edge. Since G is a cycle, there exists $w \in \sigma^*$ such that uw is $\alpha-$strong. Hence the path wuv is a unique path in G. Removal of uv affects the connectivity between w and v. Also, any two $\beta-$strong edges in G form a fuzzy edge cut of G. Thus the result follows. ■

For even saturated fuzzy cycles, we have a better situation.

Theorem 9.1.8 *Let $G = (\sigma, \mu)$ be an even saturated fuzzy cycle with $|\sigma^*| \geq 3$. Then $\kappa'(G) = 2\mu(xy)$, where xy is a weakest edge of G.*

Proof Let $G = (\sigma, \mu)$ be an even saturated fuzzy cycle with $|\sigma^*| \geq 3$. Then every vertex in G is adjacent to at least one β-strong edge. No single edge will form a fuzzy edge cut. Collection of any two edges will form a minimal FEC of which two edges with minimum weights constitute a minimum fuzzy edge cut. Hence $\kappa'(G) = 2\mu(xy)$. ■

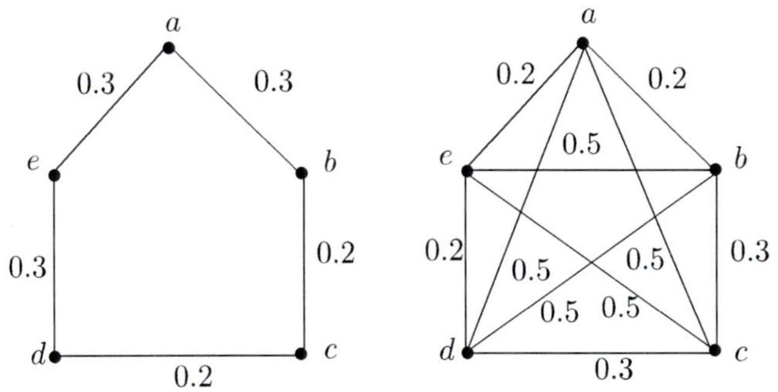

(*a*) Fuzzy graph G (*b*) G^c

Fig. 9.2 Fuzzy graphs in Example 8.1.10

Theorem 9.1.9 *Let* $G = (\sigma, \mu)$ *be a fuzzy cycle with* $|\sigma^*| = n \geq 5$ *and* $\sigma(u) = t$*,where* $t \in (0, 1]$*. Then the fuzzy edge connectivity of* $G^c = (\sigma, \nu)$ *is* $\kappa'(G^c) = (n-3)t$*.*

Proof Let $G = (\sigma, \mu)$ be a fuzzy cycle with $|\sigma^*| = n \geq 5$. Let $\sigma(u) = t$, $t \in (0, 1]$. Being a fuzzy cycle, G contains only strong edges. Also, $CONN_{G^c}(x, y) = t$, $\forall x, y \in \sigma^*$. Clearly $uv \in \mu^*$ is a $\delta-$edge in G^c, and there are exactly $n-3$ strong edges incident at every vertex in G^c. Thus the removal of a set S of exactly $n-3$ vertices from G results in the inequality $CONN_{G-S}(x, y) < CONN_G(x, y)$ $\forall x, y \in \sigma^*$. Clearly, there exists a pair of vertices u, v with $uv \in \mu^*$ such that $S(P) = t$ for every $u - v$ path in G^c. Thus by definition $\kappa'(G^c) = (n-3)t$. ■

Example 9.1.10 provides an illustration of Theorem 9.1.9.

Example 9.1.10 Consider the fuzzy graph $G = (\sigma, \mu)$ given in Fig. 9.2 with $V = \{a, b, c, d, e\}$. Define fuzzy subsets σ of V and μ of $E = \{ab, bc, cd, de, ea\}$ as follows. Let $\sigma(x) = 0.5$ for every $x \in \sigma^*$, $\mu(ab) = \mu(ae) = \mu(ed) = 0.3$, $\mu(bc) = 0.2 = \mu(cd)$.

In this example, G is a fuzzy cycle with value of t in Theorem 9.1.9 is 0.5. Consider the fuzzy graph G^c. Here, $\mu(cd) = 0.3$ and $S(P) = 0.5$ for all other strong paths from c to d. There are exactly two strong paths from c to d. Also, ce and ac are the strong edges incident at c. Thus $S = \{ce, ac\}$ is a minimum fuzzy cut set for G^c. Hence, $\kappa'(G^c) = 1$.

In the next result we compare the fuzzy edge connectivity of a fuzzy graph and the complete fuzzy graph spanned on the same vertex set.

Theorem 9.1.11 *Let* $G = (\sigma, \mu)$ *be a connected fuzzy graph with* $|\sigma^*| = n$*. Then,* $\kappa'(G) \leq \kappa'(G')$*, where* $G' = (\sigma', \mu')$ *is the complete fuzzy graph spanned by the vertex set of* G*.*

Proof Let $G = (\sigma, \mu)$ be a connected fuzzy graph and $G' = (\sigma', \mu')$ be the complete fuzzy graph spanned by σ. Let $v \in \sigma^*$ be a vertex of minimum degree in G'. Then $\kappa'(G') = (n-1)\sigma(v)$. Let E denote the set of edges incident at v in G' and E' those in G. Then E' is a fuzzy edge cut in G and hence $s(E') \le s(E) = \kappa'(G')$. Therefore, the conclusion holds for G' also. ∎

Theorem 9.1.12 *Let $G = (\sigma, \mu)$ be a fuzzy tree and F be the corresponding MST of G. Then $\kappa'(G) = \kappa'(F)$.*

The proof of Theorem 9.1.12 follows from the fact that every strong edge of G belongs to F.

9.2 Generalized Fuzzy Vertex and Edge Connectivity

Most of the definitions of vertex cut and edge cut in fuzzy graphs were in line with classical cuts, providing a disconnection of the fuzzy graph. A modification came in 2010, but it was a strong edge version. This section discusses a proper generalization of cuts, appeared in [151] in 2020. In the 2010 version by Mathew and Sunitha, only strong edges were considered. A fuzzy edge cut was defined to be a collection of strong edges, $E = \{e_1, e_2, \cdots, e_n\}$ with either $CONN_{G-E}(x, y) < CONN_G(x, y)$ for some pair of vertices $x, y \in \sigma^*$ with at least one of x or y different from both u_i and v_i, $i = 1, 2, \cdots, n$, or $G - E$ is disconnected. But even non strong edges contribute towards the connectivity of fuzzy graphs significantly. Also, the condition that x and y should be different from the end vertices of edges in E disqualifies a fuzzy edge cut, being a generalization of classical cut set of edges. The modified definitions are given in Definitions 9.2.1 and 9.2.2.

Definition 9.2.1 A set of vertices $X = \{v_1, v_2, \ldots, v_m\} \subset \sigma^*$ of a connected fuzzy graph $G = (\sigma, \mu)$ is said to be a **generalized fuzzy vertex cut**, if either, $CONN_{G-X}(x, y) < CONN_G(x, y)$ for some pair of vertices $x, y \in \sigma^* \setminus X$ or $G - X$ is trivial. The **weight** of X, denoted by $w_f(X)$ is defined as $w_f(X) = \sum_{x \in X} \mu(xy)$, where $\mu(xy)$ is the minimum of the weights of edges incident at x. The **generalized fuzzy vertex connectivity** of a connected fuzzy graph G is defined as the minimum weight of a generalized fuzzy vertex cut in G and is denoted by $\kappa_f(G)$.

A generalized fuzzy vertex cut is also called a generalized fuzzy node cut (abbreviated $g - FNC$) occasionally. Similar generalized definition for edges is as follows.

Definition 9.2.2 Let $G = (\sigma, \mu)$ be a fuzzy graph. A set of edges $E = \{e_1, e_2, \ldots, e_n\}$ is said to be a **generalized fuzzy edge cut** ($g - FEC$) if either $CONN_{G-E}(x, y) < CONN_G(x, y)$ for some pair of vertices $x, y \in \sigma^*$ or $G - E$ is disconnected.

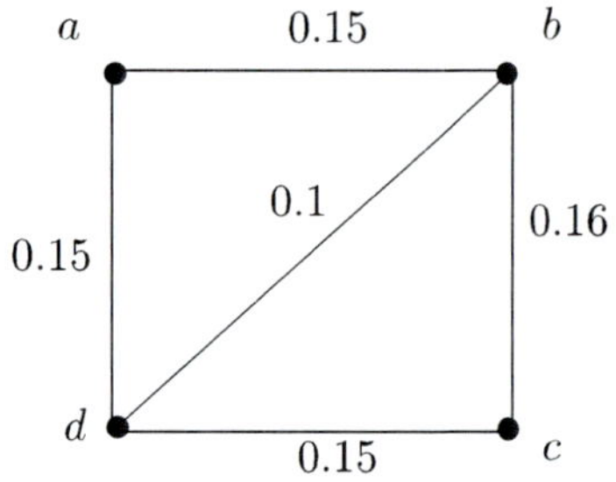

Fig. 9.3 A fuzzy graph with $\kappa_f(G) = 0.2$ and $\kappa'_f(G) = 0.16$

The **weight** of E is defined as $w'_f(E) = \sum_{e_i \in E} \mu(e_i)$. The **generalized fuzzy edge connectivity**, $\kappa'_f(G)$, of a connected fuzzy graph G is the minimum weight of a generalized fuzzy edge cut in G.

When the fuzzy graph G is disconnected, we define, $\kappa_f(G) = \kappa'_f(G) = 0$. The fuzzy graph in Example 9.2.3 illustrates the above definitions.

Example 9.2.3 Let $G = (\sigma, \mu)$ be the fuzzy graph given in Fig. 9.3 with $\sigma^* = \{a, b, c, d\}$. Here, $\mu^* = \{ab, bc, cd, ad, bd\}$, with $\sigma(x) = 1 \, \forall \, x \in \sigma^*$ and $\mu(ab) = 0.15$, $\mu(bc) = 0.16$, $\mu(cd) = 0.15$, $\mu(ad) = 0.15$, and $\mu(bd) = 0.1$. In this fuzzy graph, we can see that $\kappa(G) = 0.3$, $\kappa_f(G) = 0.2$, $\kappa'(G) = 0.3$ and $\kappa'_f(G) = 0.16$.

As expected, we can see that the new parameters give lesser values than the old ones as seen from Theorem 9.2.4.

Theorem 9.2.4 *Let G be a connected fuzzy graph. Then* $\kappa_f(G) \leq \kappa(G)$ *and* $\kappa'_f(G) \leq \kappa'(G)$.

Proof Let $G = (\sigma, \mu)$ be a connected fuzzy graph, where $\sigma^* = \{v_1, v_2, \cdots, v_n\}$. Let $X = \{v_1, v_2, \ldots, v_m\}$ be a fuzzy vertex cut with $\kappa(G) = s(X)$. Then by definition, the removal of X from G reduces the strength of connectedness between some pair of vertices $x, y \in \sigma^* \setminus X$. i.e., $CONN_{G \setminus X}(x, y) < CONN_G(x, y)$. Since we have $w_f(X) \leq s(X)$, it follows that $\kappa_f(G) = \wedge\{w_f(B) : B \, is \, a \, g - FNC \, of \, G\} \leq w_f(X) \leq s(X) = \kappa(G)$.

Let $E = \{e_1, e_2, \ldots, e_n\}$ be a minimum FEC in G. Then $\kappa'(G) = s'(E)$. Also, E satisfies all the conditions for a $g - FEC$, and hence, $\kappa'_f(G) \leq \kappa'(G)$. ■

We have seen the relationship between vertex connectivity of a fuzzy graph and that of its subgraphs in Theorem 9.1.1. The below example in Fig. 9.4 shows that this result is not true in the case of a $g - FNC$.

In Fig. 9.4, (a) $\kappa(G) = 0.3$ and $\kappa_f(G) = 0.2$. (b) $\kappa(H) = 0.3$ and $\kappa_f(H) = 0.3$. (c) $\kappa(H') = 0.15$ and $\kappa_f(H') = 0.1$. Thus, from (a) and (b), we see that, $\kappa_f(H') \leq \kappa_f(G) \leq \kappa_f(H)$. Thus, for a connected fuzzy graph G with a partial fuzzy subgraph

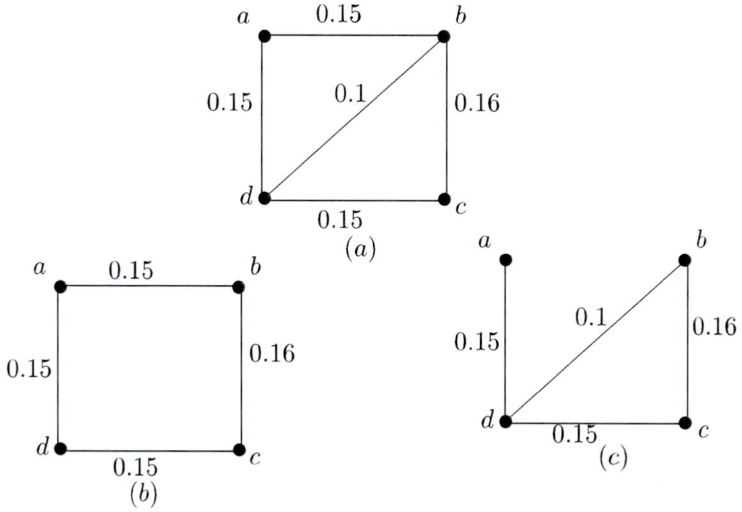

Fig. 9.4 Fuzzy graph G in (**a**) and its subgraphs (**b**) and (**c**)

H on the same vertex set, it is not generally true that $\kappa_f(H) \leq \kappa_f(G)$. Also it can be seen that $\kappa'(G)$ is not always less than or equal to $\lambda(G)$ in general, but it is always true that $\kappa'_f(G) \leq \lambda(G)$ as can be seen from Theorem 9.2.5.

Theorem 9.2.5 *Let G be a connected fuzzy graph. Then* $\kappa'_f(G) \leq \lambda(G)$.

Proof Let $G = (\sigma, \mu)$ be a connected fuzzy graph with edge connectivity $\lambda(G)$. Let $E = (V_1, V_2)$ be a cut-set in G with weight $\lambda(G)$. Since E partitions the vertex set into two disjoint sets namely V_1 and V_2, the removal of E from G disconnects G. Let $x \in V_1$ and $y \in V_2$. Then $CONN_{G-E}(x, y) = 0 < CONN_G(x, y)$. Hence, E is a $g - FEC$ in G. Since κ'_f is the minimum weight of $g - FEC$'s in G, it follows that $\kappa'_f(G) \leq w_f(E) = \lambda(G)$. ∎

Combining Theorem 9.2.4 and Theorem 9.2.5, we have the following result.

Theorem 9.2.6 *Let* $G = (\sigma, \mu)$ *be a connected fuzzy graph. Then* $\kappa'_f(G) \leq min \{\kappa'(G), \lambda(G)\}$.

The existing fuzzy vertex connectivity and fuzzy edge connectivity are related as follows. In a connected fuzzy graph $G = (\sigma, \mu)$, $\kappa(G) \leq \kappa'(G)$. We cannot make a general relationship between $\kappa_f(G)$ and $\kappa'_f(G)$. This is illustrated in the following example.

Example 9.2.7 Let $G = (\sigma, \mu)$ be the complete fuzzy graph given in Fig. 9.5 with $\sigma^* = \{a, b, c, d\}$ and $\mu^* = \{ab, bc, cd, ad, bd\}$. Let $\sigma(a) = 0.4$, $\sigma(b) = 0.3$, $\sigma(c) = 0.3$, $\sigma(d) = 0.2$ and $\mu(ab) = 0.3$, $\mu(bc) = 0.2$, $\mu(ac) = 0.2$, $\mu(ad) = 0.2$, $\mu(bd) = 0.2$ and $\mu(cd) = 0.1$. We can see that $\kappa_f(G) = 0.6$, and $\kappa'_f(G) = 0.3$.

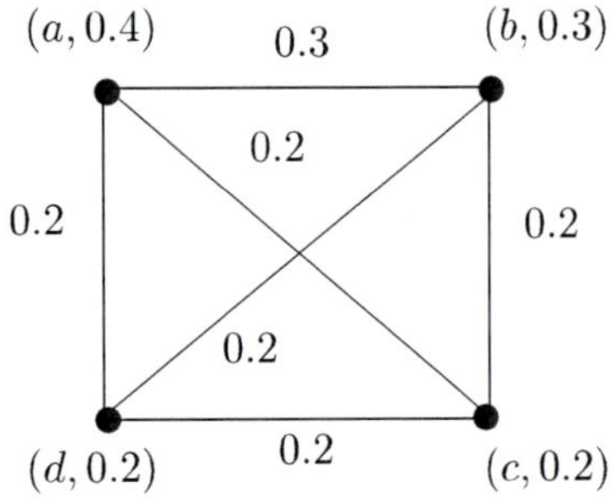

Fig. 9.5 A fuzzy graph with $\kappa_f(G) > \kappa'_f(G)$

The next theorem provides the condition for $\kappa_f(G) \leq \kappa'_f(G)$ to hold.

Theorem 9.2.8 *Let G be a fuzzy graph. If any minimum generalized fuzzy edge cut of G contains more than one element, then* $\kappa_f(G) \leq \kappa'_f(G)$.

Proof Let $G = (\sigma, \mu)$ be a connected fuzzy graph. Let E be a minimum generalized fuzzy edge cut. Then $w'_f(E) = \kappa'_f$. Clearly, E does not contain α-strong edges, for otherwise, it will contradict the minimality of E. Consider $G - E$. Then there exist $u, v \in \sigma^*$ such that $CONN_{G-E}(u, v) < CONN_G(u, v)$. Let S be the set of end vertices of edges in E. If $u, v \notin S$, then E is a fuzzy edge cut and hence $\kappa' \leq \kappa'_f$. However, $\kappa'_f \leq \kappa'$. Therefore, $\kappa'_f = \kappa'$.

Similarly, we can prove that $\kappa_f(G) = \kappa(G)$. Hence it follows that, $\kappa_f(G) \leq \kappa'_f(G)$. Either u or v belongs to S. Let K be the set of vertices in S which are adjacent either to u or to v. Then $CONN_{G-K}(u, v) < CONN_G(u, v)$. Therefore, K becomes a $g - FNC$ and hence, $\kappa'_f = \kappa'$. Hence the theorem is proved. If G is not connected, then the theorem is trivially true. ■

Corollary 9.2.9 *If a connected fuzzy graph G contains only* $\beta-$*strong edges and* $\delta-$*edges, then* $\kappa_f(G) \leq \kappa'_f(G)$.

Proof Let $G = (\sigma, \mu)$ be a connected fuzzy graph which contain only $\beta-$strong and $\delta-$edges. Then any minimum $g - FEC$ of G contains more than one element. Thus, the proof follows by Theorem 9.2.8. ■

It is difficult to say anything about the relationship between $g - FNC$ and $g - FEC$ in a connected fuzzy graph G, having a minimum $g - FEC$ containing $\alpha-$strong edges. This is shown in Fig. 9.6.

In Fig. 9.6, G is a fuzzy tree and hence $\kappa'_f(G) = 0.3$ and $\kappa_f(G) = 0.3$. H is a complete fuzzy graph with $\kappa'_f(H) = 0.3$ and $\kappa_f(H) = 0.6$.

Corollary 9.2.10 *If a fuzzy graph* $G = (\sigma, \mu)$ *satisfies* $\kappa'_f = \kappa'$, *then* $\kappa_f(G) \leq \kappa'_f(G)$.

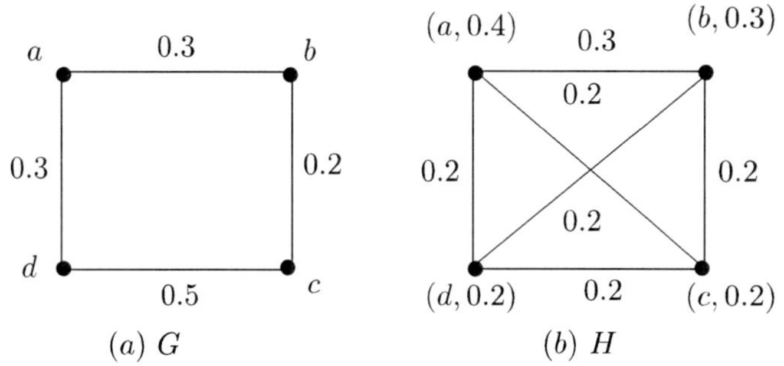

Fig. 9.6 A fuzzy tree and a complete fuzzy graph

Theorem 9.2.11 *Let $G = (\sigma, \mu)$ be a complete fuzzy graph with $\sigma^* = \{v_1, v_2, \cdots, v_m\}$. Let $\sigma(v_i) = t_i$ for $i = 1, 2, \cdots, m$ and $t_1 < t_2 < \cdots < t_m$. Then $\kappa_f(G) = (m-1)t_1$ and $\kappa'_f(G) = \wedge\{(m-1)*t_1, (m-2)*t_2, (m-3)*t_3, \cdots, 2*t_{m-2}, t_{m-1}\}$, where $*$ denote the ordinary product.*

Proof The first part follows from the fact that any collection of $m-1$ vertices form a minimal $g-FNC$ with weight $(m-1)t_1$. For the second part, $CONN_G(v_i, v_j) = t_i$ as the $v_i's$ are arranged in order and G is complete. To reduce $CONN_G(v_i, v_j)$, we must delete a set E of $m-i$ edges incident at v_i with weight t_i. Thus $w'_f(E) = (m-i)*t_i$. Since $|\sigma^*| = n$, there are $\binom{n}{2}$ $g-$fuzzy edge cuts in G. However, they can have weights from the set $\{(m-1)*t_1, (m-2)*t_2, (m-3)*t_3, \cdots, 2*t_{m-2}, t_{m-1}\}$ and so the result follows. ■

Theorem 9.2.12 *For any connected fuzzy graph with generalized fuzzy edge connectivity $\kappa'_f(G)$, there exists a connected super fuzzy graph G' of G with fuzzy edge connectivity $\kappa'_f(G') > \kappa'_f(G)$.*

Proof Suppose that G is trivial is fuzzy graph, i.e., $\sigma^* = \{u\}$, $\kappa'_f(G) = 0$. Define the super fuzzy graph $G' = (\rho, \nu)$ as follows: $\rho^* = \sigma^* \cup \{v\}$ and $\nu^* = \{uv\}$, where

$$\rho(x) = \begin{cases} \sigma(x) & if\ x = u \\ 1 & if\ x = v \end{cases}$$

and $\nu(uv) = t,\ t \in (0, 1]$. Clearly $\kappa'_f(G') > \kappa'_f(G)$.

Suppose that G is a single edge fuzzy graph, i.e., $\sigma^* = \{u, v\}$, and $\mu^* = \{uv\}$. Thus in this case $\kappa'_f(G) = \mu(uv) = t$(say). Construct a super graph $G' = (\rho, \nu)$ with $\rho^* = \sigma^* \cup \{z\}$ and $\nu^* = \mu^* \cup \{uz, vz\}$, where

$$\rho(x) = \begin{cases} \sigma(x) & if\ x \in \sigma^* \\ 1 & if\ x = z \end{cases}$$

and

$$\nu(xy) = \begin{cases} \mu(xy) & if\ xy = uv \\ s & if\ xy = uz\ xy = vz,\ s \nleq t. \end{cases}$$

If $s > t$, Then, $\kappa'_f(G') = s > t = \kappa'_f(G)$, else then, $\kappa'_f(G') = 2s > t = \kappa'_f(G)$.

Suppose that G is a fuzzy graph with n vertices where $n > 2$, say $\{v_1, v_2, \ldots, v_n\}$, $n \geq 3$. Let $E_1, E_2, \cdots, E_k$ be minimum generalized fuzzy edge cuts of G and $X_1, X_2, \cdots, X_k$ be its end vertices corresponding to each E_j, $j \leq k$. Construct super graphs $G_1, G_2, \cdots, G_k$ with $G_k = (\rho, \nu)$, where $\rho^* = \sigma^* \cup \{z\}$ and $\nu^* = \mu^* \cup \{uz : u \in X_k\}$,

$$\rho(x) = \begin{cases} \sigma(x) & if\ x \in \sigma^* \\ \wedge\{\sigma(x) : x \in X_k\} & if\ x = z \end{cases}$$

and

$$\nu(xy) = \begin{cases} \mu(xy) & if\ xy \in \mu^* \\ \sigma(z) & otherwise. \end{cases}$$

Thus, every G_j $j \geq k$, we have $\kappa'_f(G_j) \geq \kappa'_f(G)$. If all $\kappa'_f(G_j) = \kappa'_f(G)$, then $G' = \bigcup_{j=1}^{k} G_j$ is super graph with $\kappa'_f(G_j) > \kappa'_f(G)$. ■

As expected the generalized edge connectivity value of a fuzzy tree coincides with the fuzzy edge connectivity as shown in the following result.

Theorem 9.2.13 *For a fuzzy tree* $G = (\sigma, \mu)$, $\kappa'_f(G) = \wedge\{\mu(xy) : xy$ *is a strong edge in* $G\}$.

Proof Let $G = (\sigma, \mu)$ be a fuzzy tree. Suppose that G has a cycle, say C. Then there exists a weakest edge xy in C, which is a δ−edge. All other edges in C are strong. Hence the removal of any edge in C other than xy reduces $CONN_G(x, y)$. An edge not belonging to a cycle in G is a bridge and hence is a $g - FEC$.

If G has no cycles, then G^* is a tree. Every edge of G is bridge and trivially becomes a $g - FEC$. Thus it follows that $\kappa'_f(G)$ equals the membership values of strong edges in G. ■

As in case of fuzzy node cuts, it is not generally true that for a fuzzy tree $\kappa_f(G) = \wedge\{$membership values of strong edges in $G\}$, as seen from the fuzzy graph

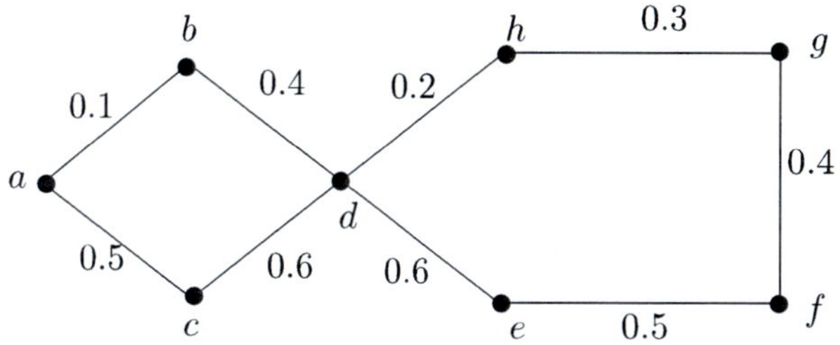

Fig. 9.7 A fuzzy tree with $\kappa_f(G) = 0.2$

in Fig. 9.7, whose generalized fuzzy vertex connectivity is 0.2, where as the minimum membership value of strong edges in G is 0.3. Theorem 9.2.14 is trivial.

Theorem 9.2.14 *For a fuzzy tree $G = (\sigma, \mu)$, $\kappa'_f(G - uv) \leq \kappa'_f(G)$ for all $uv \in \mu^*$.*

Theorem 9.2.15 *Let $G = (\sigma, \mu)$ be a fuzzy tree and let $d_s(u) = d_s(v) = 1$ for every $\delta-$edge uv in G. Then $\kappa_f(G) = \wedge\{\mu(xy) : xy$ is a strong edge in $G\}$.*

Proof It follows that there exists no $g - FNC$ containing the end vertices of $\delta-$edges alone. Consider all strong edges in G. Let $e = xy$ be a strong edge with minimum $\mu-$value. Then either $S_1 = \{x\}$ or $S_2 = \{y\}$ becomes a minimum $g - FNC$. In fact, S_2 becomes a $g - FNC$ when $d_\alpha(y) = 2$ and S_1 becomes a $g - FNC$ when $d_\alpha(x) = 2$. Thus $\kappa_f(G) = \wedge\{\mu(xy) : xy$ is a strong edge in $G\}$. ■

Corollary 9.2.16 *Let $G = (\sigma, \mu)$ be a fuzzy tree such that G^* is a cycle, then $\kappa_f(G) = \wedge\{\mu(xy) : xy$ is a strong edge in $G\}$.*

The next two theorems speaks about the existence of a generalized $t-$connected and $t-$edge connected complete fuzzy graph for any real t.

Theorem 9.2.17 *There exists a generalized $t-$connected complete fuzzy graph for any $t \in \mathbb{R}^+$.*

Proof Let us first assume that $t \in (0, 1]$. A $t-$connected complete fuzzy graph is simply $G = K_2$ with $\sigma^* = \{u, v\}$ and $\mu(uv) \geq t$.

Suppose $t \geq 1$. Let $t \in [n, n + 1)$ for some $n \in \mathbb{N}$. Construct a fuzzy graph G with at least $n + 2$ vertices namely $v_1, v_2, \ldots, v_{n+2}, \ldots, v_{n+k}$. Let $\sigma(v_i) \geq \dfrac{t}{(n + k)}$ for $i = 1, 2, \ldots, n + k$, along with $\mu(v_i v_j) = \sigma(v_i) \wedge \sigma(v_j)$ for $i \neq j$. Then G is a CFG with $n + k$ vertices. Without loss of generality we can assume v_1 be the vertex of minimum degree. Thus by Theorem 8.2.11, $\kappa_f(G) = (n + k) * \wedge\{\sigma(x) : x \in \sigma^*\}$. Thus, $\kappa_f(G) \geq (n + k) * \dfrac{t}{(n + k)} = t$. Thus, G is a generalized $t-$connected complete fuzzy graph. ■

Theorem 9.2.18 *There exists a generalized $t-$edge connected complete fuzzy graph for any $t \in \mathbb{R}^+$.*

Proof Let us first assume that $t \in (0, 1]$. A generalized $t-$connected complete fuzzy graph is simply $G = K_2$ with $\sigma^* = \{u, v\}$ and $\mu(uv) \geq t$.

Suppose $t \geq 1$. Let $t \in [n, n+1)$ for some $n \in \mathbb{N}$. Construct a fuzzy graph G with at least $n+2$ vertices namely $v_1, v_2, \ldots, v_{n+2}, \ldots, v_{n+k}$. Let $\sigma(v_i) \geq \frac{t}{(n+k)-i}$ for $i = 1, 2, \ldots, k, k+1$ be such that no v_i gets the same value. Let $p = \wedge\{\sigma(v_i) : i = 1, 2, \ldots, k+1\}$, $q = \vee\{\sigma(v_i) : i = 1, 2, \ldots, k+1\}$ and $\sigma(v_i) = q \ \forall v_i$ for $i = k+1, , k+2, \ldots, k+n$. Let $\mu(v_i v_j) = \sigma(v_i) \wedge \sigma(v_j)$ for $i \neq j$. By construction, G is a complete fuzzy graph. Arrange the vertices in an ascending order of their weights and rename as $u_1, u_2, \ldots, u_{n+k}$, $\sigma(u_i) = t_i$ for $i = 1, 2, \cdots, n+k$ and $t_1 < t_2 < \cdots < t_{k+1}$. Without loss of generality we can use the Theorem 9.2.11 with $n+k$ vertices and $\kappa'_f(G) = \wedge\{(n+k-1) * t_1, (n+k-2) * t_2, (n+k-3) * t_3, \cdots, (n+1) * t_{k-1}, (n) * t_k\}$. Since $(n+k-j) * t_j = (n+k-j) * \sigma(u_j) = (n+k-j) * \sigma(v_l) \geq (n+k-j) * \frac{t}{(n+k)-l}$. Clearly, it can be seen that $(n+k-j) * t_j \geq t$ for any j, as j varies only from $1, 2, \ldots, k$. Thus by definition $\kappa'_f(G) \geq t$. ■

Theorem 9.2.19 *Let $G = (\sigma, \mu)$ be a fuzzy cycle. Then $\kappa'_f(G) = \wedge\{m, 2\mu(xy)\}$, where xy is a weakest edge in G, and m is the minimum membership value of $\alpha-$strong edges in G.*

Proof Let $G = (\sigma, \mu)$ be a fuzzy cycle. Then all edges in G are strong. Clearly, any set consisting of a single $\alpha-$strong edge or two $\beta-$strong edges constitutes $g - FEC$'s in G. Let m be the minimum membership value of $\alpha-$strong edges in G and let xy be an edge of minimum $\mu-$value. Note that any $\beta-$strong edge in G will have membership value $\mu(xy)$ and any $g - FEC$ containing two $\beta-$strong edges has strength equal to $2\mu(xy)$. Thus, $\kappa'_f(G) = \wedge\{m, 2\mu(xy)\}$. ■

Theorem 9.2.20 *Let $G = (\sigma, \mu)$ be a β-saturated fuzzy cycle. Then $\kappa'_f(G - v) = \kappa'_f(G - uv) = k \ \forall v \in \sigma^*, \ uv \in \mu^*$, where k is the membership value of weakest edge in G.*

Proof Suppose G is a β-saturated fuzzy cycle. Then every vertex is adjacent to at least one β-strong edge. Clearly, the removal of a vertex from G results in a tree. Since G is $\beta-$saturated, $G - u$ has at least one edge say uv such that $\mu(uv) = k$, where k is the minimum membership value of weakest edge in G. Then by Theorem 8.2.5, the result follows. ■

From the fuzzy graph shown in Fig. 9.8 we can see that this result is true is not true in general. Note that fuzzy graph in Fig. 9.8 is not $\beta-$saturated and Theorem 9.2.20 fails.

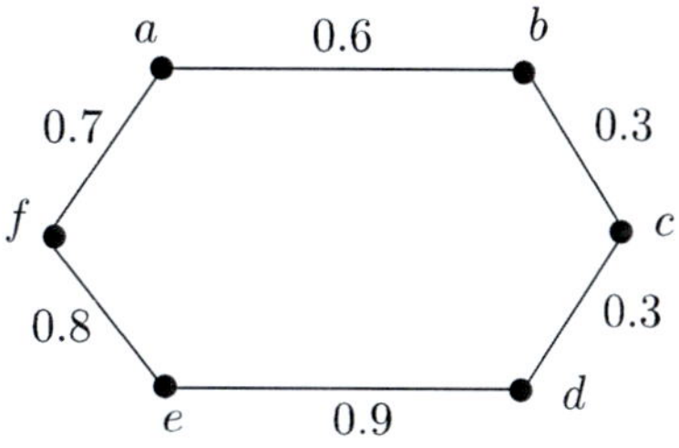

Fig. 9.8 Fuzzy graph G with $\kappa'_f(G) = 0.6$

For the fuzzy graph in Fig. 9.8, $\kappa'_f(G) = 0.6$. However, $\kappa'_f(G - b) = 0.3$. G does not satisfy Theorem 9.2.20.

Theorem 9.2.21 *Let $G = (\sigma, \mu)$ be a fuzzy cycle with $|\sigma^*| = n \geq 5$ and $\sigma(u) = t$, $t \in (0, 1]$. Then for $G^c = (\sigma, \nu)$, $\kappa_f(G^c) \leq (n - 3) * \vee\{t - \mu(uv) : \mu(uv) \neq 0\}$.*

Proof Let $G = (\sigma, \mu)$ be a fuzzy cycle with $|\sigma^*| = n \geq 5$ and $\sigma(u) = t$, $t \in (0, 1]$. Since G is a fuzzy cycle, G contains only strong edges. Also, $CONN_{G^c}(x, y) = t \; \forall x, y \in \sigma^*$. Clearly, any $uv \in \mu^*$ becomes a $\delta-$edge in G^c and there are exactly $n - 3$ strong edges incident at every vertex in G^c. Thus any $g - FNC$ X in G^c has cardinality $n - 3$. Also each vertex in X is adjacent with at least one $\delta-$edge whose $\nu-$value is strictly less than k. Let m be the minimum membership value of edges in G. Then $\nu(xy) \leq t - m \; \forall \; xy \in \mu^*$. Hence $\kappa_f(G^c) \leq (n - 3) * \vee\{t - \mu(uv) : \mu(uv) \neq 0\}$. ■

We next provide a lower bound for the generalized fuzzy vertex connectivity of complements of fuzzy cycles as in the following theorem. The proof is similar and is omitted.

Theorem 9.2.22 *Let $G = (\sigma, \mu)$ be a fuzzy cycle with $|\sigma^*| = n \geq 5$ and $\sigma(u) = t$, $t \in (0, 1]$. Then the generalized fuzzy vertex connectivity of $G^c = (\sigma, \nu)$ satisfies $\kappa_f(G^c) \geq (n - 3) * \wedge\{t - \mu(uv) : \mu(uv) \neq 0\}$.*

Definition 9.2.23 Let G be a connected fuzzy graph and $t \in (0, \infty)$. Then G is called **generalized t-fuzzy connected** if $\kappa_f(G) \geq t$ and G is called **generalized t-fuzzy edge connected** if $\kappa'_f(G) \geq t$.

Speaking differently, a fuzzy graph G is t-fuzzy connected if there exist no generalized fuzzy vertex cut with weight less than t and is t-fuzzy edge connected if there exist no generalized fuzzy edge cut with weight less than t.

Theorem 9.2.24 *Every generalized $t-$fuzzy connected graph G is a $t-$connected fuzzy graph.*

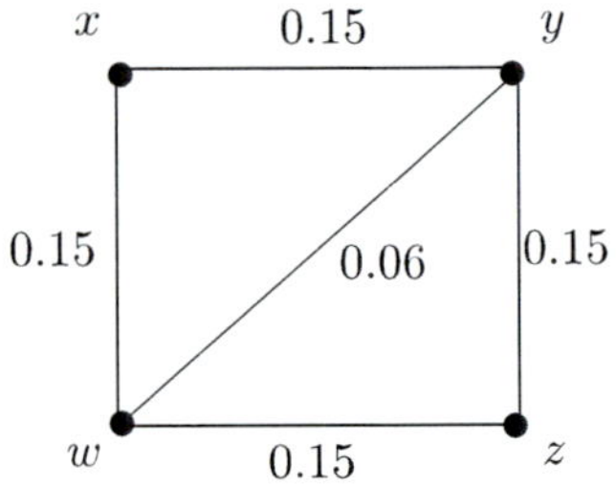

Fig. 9.9 Fuzzy graph with $\kappa(g) = 0.3$ and $\kappa_f(G) = 0.12$.

Proof Let G be a generalized $t-$fuzzy connected graph, so, $\kappa_f(G) \geq t$. Then by Theorem 9.2.4, we have $t \leq \kappa(G)$. Thus G is $t-$connected. ■

Below Fig. 9.9 shows that the converse of the above theorem need not true.

The fuzzy graph G in Fig. 9.9 is $0.29-$connected, but it is not generalized $0.29-$fuzzy connected.

9.3 Fuzzy Graph Clustering

In this section, we focus on algorithms for clustering of fuzzy graphs, based on the different connectivity parameters. The algorithm based on generalized connectivity seem to be better than the ones based on other algorithms for fuzzy graph clustering. Both techniques are compared. Some of the definitions and results, used in the process are given below.

Theorem 9.3.1 *A generalized $t-$fuzzy edge component of $G = (\sigma, \mu)$ is a maximal generalized $t-$fuzzy edge connected fuzzy subgraph of $G = (\sigma, \mu)$. In other words a maximal generalized $t-$fuzzy edge connected subgraph, is a fuzzy subgraph H of G, induced by a set of vertices in G such that $\kappa'_f(H) = t$.*

The above concept is illustrated in the following example.

Example 9.3.2 Consider the fuzzy graph G with $\sigma^* = \{a, b, c, d\}$ and $\sigma(x) = 1 \; \forall x \in \sigma^*$, $\mu(ab) = \mu(cd) = 0.3$, $\mu(bc) = 0.7$, and $\mu(ad) = 0.8$.

Here, $\kappa'_f(G) = 0.6$. Hence, G is generalized $t-$fuzzy edge connected for all t such that $t \leq 0.6$. Thus, G itself is a generalized $t-$fuzzy edge component for all t such that $0 \leq t \leq 0.6$. Now let $t = 0.7$. Then the generalized $0.7-$fuzzy edge components of G are $H_1 = \{b, c\}$, $H_2 = \{a\}$, $H_3 = \{d\}$ with $\kappa'_f(H_1) = 0.7$. When $t = 0.8$, the generalized fuzzy edge components of G are $H_4 = \{a, d\}$, $H_5 = \{b\}$, $H_6 = \{c\}$ with $\kappa'_f(H_4) = 0.8$.

Definition 9.3.3 Let $G = (\sigma, \mu)$ be a fuzzy graph. A collection C of vertices in G is called a generalized fuzzy cluster of level t if the fuzzy subgraph of G induced by C is a generalized $t-$fuzzy edge component of G.

The concept of a cohesive matrix M was defined by Yeh and Bang [171]; we use it to find the maximal $t-$edge connected components of a fuzzy graph G. An element of G is defined to be either a vertex or an edge. The cohesiveness of an element denoted by $h(e)$, is the maximum value of edge connectivity of the subgraphs of G containing e. The cohesive matrix M of G is defined as $M = (m_{i,j})$, where $m_{i,j} =$ the cohesiveness of the edge $v_i v_j$ if $i \neq j$ and the cohesiveness of the vertex v_i if $i = j$. The definition of a generalized cohesive matrix is given below.

Definition 9.3.4 Let $G = (\sigma, \mu)$ be a fuzzy graph. An element of G is defined to be either a vertex or an edge. The generalized cohesiveness of an element denoted by $h(e)$, is the maximum value of generalized edge connectivity of the subgraphs of G containing e.

Definition 9.3.5 Let $G = (\sigma, \mu)$ be a fuzzy graph. The generalized cohesive matrix M of G is defined as $M = (m_{i,j})$, where $m_{i,j}$ is the generalized cohesiveness of the edge $v_i v_j$ if $i \neq j$ and the generalized cohesiveness of the vertex v_i if $i = j$.

The generalized cohesive matrix M can be used to find the maximal generalized $t-$fuzzy edge connected components of a fuzzy graph G.

We say that vertex $v \in \sigma^*$ is said to be in a cluster of level t, if v belongs to a generalized $t-$fuzzy edge component of G. Thus, finding the generalized $t-$fuzzy edge components of G is equivalent to the extraction of clusters from G. This process of finding generalized $t-$fuzzy edge components and fuzzy clusters using generalized fuzzy edge connectivity is termed as *generalized t-fuzzy edge connectivity procedure*.

Generalized t-Fuzzy Edge Connectivity Procedure:

Step 1: Obtain the generalized cohesive matrix M of the fuzzy graph G.
Step 2: Obtain the t-threshold graph G_t of M.
Step 3: The maximal complete subgraphs of G_t are the generalized $t-$fuzzy edge components.

Illustration: Consider the following part of the grid network represented as a fuzzy subgraph given in Fig. 9.10.

The matrix representation of the fuzzy graph in Fig. 9.10 with 10 nodes is,

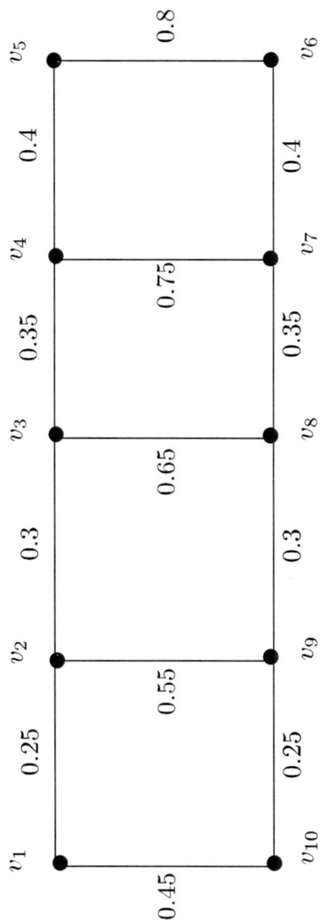

Fig. 9.10 A grid network

$$G = \begin{bmatrix} 0 & 0.25 & 0 & 0 & 0 & 0.45 & 0 & 0 & 0 & 0 \\ 0.25 & 0 & 0.3 & 0 & 0 & 0 & 0.55 & 0 & 0 & 0 \\ 0 & 0.3 & 0 & 0.35 & 0 & 0 & 0 & 0.65 & 0 & 0 \\ 0 & 0 & 0.35 & 0 & 0.4 & 0 & 0 & 0 & 0.75 & 0 \\ 0 & 0 & 0 & 0.4 & 0 & 0 & 0 & 0 & 0 & 0.8 \\ 0.45 & 0 & 0 & 0 & 0 & 0 & 0.25 & 0 & 0 & 0 \\ 0 & 0.55 & 0 & 0 & 0 & 0.25 & 0 & 0.3 & 0 & 0 \\ 0 & 0 & 0.65 & 0 & 0 & 0 & 0.3 & 0 & 0.35 & 0 \\ 0 & 0 & 0 & 0.75 & 0 & 0 & 0 & 0.35 & 0 & 0.4 \\ 0 & 0 & 0 & 0 & 0.8 & 0 & 0 & 0 & 0.4 & 0 \end{bmatrix}$$

The cohesive matrix M of the fuzzy graph is,

$$M = \begin{bmatrix} 0 & 0.5 & 0.5 & 0.5 & 0.5 & 0.5 & 0.5 & 0.5 & 0.5 & 0.5 \\ 0.5 & 0 & 0.6 & 0.6 & 0.6 & 0.5 & 0.6 & 0.6 & 0.6 & 0.6 \\ 0.5 & 0.6 & 0 & 0.7 & 0.7 & 0.5 & 0.6 & 0.7 & 0.7 & 0.7 \\ 0.5 & 0.6 & 0.7 & 0 & 0.8 & 0.5 & 0.6 & 0.7 & 0.8 & 0.8 \\ 0.5 & 0.6 & 0.7 & 0.8 & 0 & 0.5 & 0.6 & 0.7 & 0.8 & 0.8 \\ 0.5 & 0.5 & 0.5 & 0.5 & 0.5 & 0 & 0.5 & 0.5 & 0.5 & 0.5 \\ 0.5 & 0.6 & 0.6 & 0.6 & 0.6 & 0.5 & 0 & 0.6 & 0.6 & 0.6 \\ 0.5 & 0.6 & 0.7 & 0.7 & 0.7 & 0.5 & 0.6 & 0 & 0.7 & 0.7 \\ 0.5 & 0.6 & 0.7 & 0.8 & 0.8 & 0.5 & 0.6 & 0.7 & 0 & 0.8 \\ 0.5 & 0.6 & 0.7 & 0.8 & 0.8 & 0.5 & 0.6 & 0.7 & 0.8 & 0 \end{bmatrix}$$

The generalized cohesive matrix M of G is given by,

$$M = \begin{bmatrix} 0 & 0.45 & 0.45 & 0.45 & 0.45 & 0.45 & 0.45 & 0.45 & 0.45 & 0.45 \\ 0.45 & 0 & 0.55 & 0.55 & 0.55 & 0.45 & 0.55 & 0.55 & 0.55 & 0.55 \\ 0.45 & 0.55 & 0 & 0.65 & 0.65 & 0.45 & 0.55 & 0.65 & 0.65 & 0.65 \\ 0.45 & 0.55 & 0.65 & 0 & 0.75 & 0.45 & 0.55 & 0.65 & 0.75 & 0.75 \\ 0.45 & 0.55 & 0.65 & 0.75 & 0 & 0.45 & 0.55 & 0.65 & 0.75 & 0.8 \\ 0.45 & 0.45 & 0.45 & 0.45 & 0.45 & 0 & 0.45 & 0.45 & 0.45 & 0.45 \\ 0.45 & 0.55 & 0.55 & 0.55 & 0.55 & 0.45 & 0 & 0.55 & 0.55 & 0.55 \\ 0.45 & 0.55 & 0.65 & 0.65 & 0.65 & 0.45 & 0.55 & 0 & 0.65 & 0.65 \\ 0.45 & 0.55 & 0.65 & 0.75 & 0.75 & 0.45 & 0.55 & 0.65 & 0 & 0.75 \\ 0.45 & 0.55 & 0.65 & 0.75 & 0.8 & 0.45 & 0.55 & 0.65 & 0.75 & 0 \end{bmatrix}$$

We consider both clustering procedures given in Ref. [98] and the one described above. We can see that the new procedure excels compared to the old.

Generalized $t-$Fuzzy Connectivity Procedure

It is not hard to see that, $K'_f(G) = 0.45$. Using generalized $t-$connectivity procedure we can obtain the generalized fuzzy clusters as follows.

Level	Generalized fuzzy clusters
$(0.8, \infty)$	$\{v_1\}, \{v_2\}, \{v_3\}, \{v_4\}, \{v_5\}, \{v_6\}, \{v_7\}, \{v_8\}, \{v_9\}, \{v_{10}\}$
$(0.75, 0.8]$	$\{v_1\}, \{v_2\}, \{v_3\}, \{v_4\}, \{v_6\}, \{v_7\}, \{v_8\}, \{v_9\}, \{v_5, v_{10}\}$
$(0.65, 0.75]$	$\{v_1\}, \{v_2\}, \{v_3\}, \{v_6\}, \{v_7\}, \{v_8\}, \{v_4, v_5, v_9, v_{10}\}$
$(0.55, 0.65]$	$\{v_1\}, \{v_2\}, \{v_6\}, \{v_7\}, \{v_3, v_4, v_5, v_8, v_9, v_{10}\}$
$(0.45, 0.55]$	$\{v_1\}, \{v_6\}, \{v_2, v_3, v_4, v_5, v_7, v_8, v_9, v_{10}\}$
$(0, 0.45]$	$\{v_1, v_2, v_3, v_4, v_5, v_6, v_7, v_8, v_9, v_{10}\}$

Consider the same fuzzy graph used for the above illustration. The clusters obtained while applying $t-$fuzzy edge connectivity procedure in Ref. [98] is given below.

$t-$Fuzzy Edge Connectivity Procedure

We have $K'(G) = 0.5$, using $t-$fuzzy edge connectivity procedure, we obtain fuzzy clusters as,

Level	Fuzzy clusters
$(0.8, \infty)$	$\{v_1\}, \{v_2\}, \{v_3\}, \{v_4\}, \{v_5\}, \{v_6\}, \{v_7\}, \{v_8\}, \{v_9\}, \{v_{10}\}$
$(0.7, 0.8]$	$\{v_1\}, \{v_2\}, \{v_3\}, \{v_6\}, \{v_7\}, \{v_8\}, \{v_4, v_5, v_9, v_{10}\}$
$(0.6, 0.7]$	$\{v_1\}, \{v_2\}, \{v_6\}, \{v_7\}, \{v_3, v_4, v_5, v_8, v_9, v_{10}\}$
$(0.5, 0.6]$	$\{v_1\}, \{v_6\}, \{v_2, v_3, v_4, v_5, v_7, v_8, v_9, v_{10}\}$
$(0, 0.5]$	$\{v_1, v_2, v_3, v_4, v_5, v_6, v_7, v_8, v_9, v_{10}\}$

We see that the generalized $t-$fuzzy edge connectivity procedure finds more qualitative clusters. For example, the cluster corresponding to the threshold value $t = 0.68$ is not present in the second set of clusters. Also note that clusters obtained through the old method appear in the new procedures for lower levels of the thresholds. This is an advantage with regard to the computational complexity involved.

Thus it follows that the generalized $t-$fuzzy edge connectivity procedure is more effective than the old $t-$fuzzy edge connectivity procedure in many ways. In the next section, this new clustering technique will be applied to human trafficking networks.

9.4 Application: Human Trafficking

Trafficking of persons is a transnational crime that is often carried out domestically or within a given subregion. Most offenders are convicted in their countries of citizenship. However, victims are often foreigners in the country where their exploitation

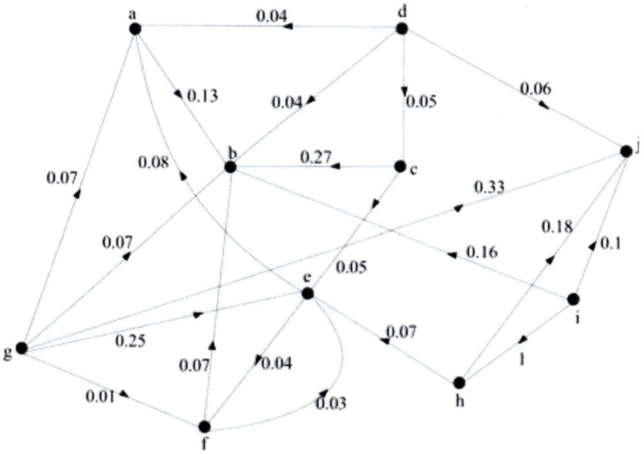

Fig. 9.11 Directed graph model of human trafficking

was detected. Trafficking flows are usually confined to a geographically limited area, either within a country or between neighboring or relatively close countries. In Chap. 2 of Ref. [58], regional overviews are given concerning trafficking situations in four regions, (1) Europe and Central Asia, (2) the Americas, (3) South Asia., East Asia and the Pacific, and (4) Africa and the Middle East. Profiles of offenders, victims, trafficking flows, and regional responses to the trafficking crime are also given.

In this section, we consider the model given in Ref. [109]. In Ref. [109], the four regions mentioned above were divided further. Human trafficking between different locations of the globe was represented as a directed fuzzy graph. Some of the most vulnerable routes are identified and major human trafficking paths are classified. The directed graph $\overrightarrow{G}$ obtained after fuzzification is given in Fig. 9.11. In this section, we apply the clustering technique developed in this chapter to the model and draw some conclusions. We use the following subregions: (a) West-Central Europe, (b) West-Southern Europe, (c) Central Europe, (d) Eastern Europe and Central Asia, (e) North and Central America and caribbeans, (f) South America, (g) East Asia and Pacific, (h) South Asia, (i) Sub Saharan Africa, and (j) Middle East.

The diagram shows the illegal flow of humans between 10 different subregions of the world. The associated connectivity matrix $A = D_f(\overrightarrow{G})$ of the directed fuzzy graph is given below.

$$A = \begin{bmatrix} 0 & 0.13 & 0 & 0 & 0 & 0 & 0 & 0 & 0 & 0 \\ 0 & 0 & 0 & 0 & 0 & 0 & 0 & 0 & 0 & 0 \\ 0.05 & 0.27 & 0 & 0 & 0.05 & 0.04 & 0 & 0 & 0 & 0 \\ 0.05 & 0.05 & 0.05 & 0 & 0.05 & 0.04 & 0 & 0 & 0 & 0.06 \\ 0.08 & 0.08 & 0 & 0 & 0 & 0.04 & 0 & 0 & 0 & 0 \\ 0.03 & 0.07 & 0 & 0 & 0.03 & 0 & 0 & 0 & 0 & 0 \\ 0.08 & 0.07 & 0 & 0 & 0.25 & 0.04 & 0 & 0 & 0 & 0.33 \\ 0.07 & 0.07 & 0 & 0 & 0.07 & 0.04 & 0 & 0 & 0 & 0.18 \\ 0.07 & 0.16 & 0 & 0 & 0.07 & 0.04 & 0 & 1 & 0 & 0.18 \\ 0 & 0 & 0 & 0 & 0 & 0 & 0 & 0 & 0 & 0 \end{bmatrix}$$

Each entry a_{ij} in A denotes the flow from ith region to jth region. In order to apply the technique developed here, the connectivity matrix A of the directed graph is reconstructed by taking $a_{ij} = a_{ji}$ as the connectivity matrix of an un-directed graph as follows. If there is no path from a to b, then the strength from b to a is assigned as $CONN_G(a, b)$.

$$B = \begin{bmatrix} 0 & 0.13 & 0.05 & 0.05 & 0.08 & 0.03 & 0.08 & 0.07 & 0.07 & 0 \\ 0.13 & 0 & 0.27 & 0.05 & 0.08 & 0.07 & 0.07 & 0.07 & 0.16 & 0 \\ 0.05 & 0.27 & 0 & 0.05 & 0.05 & 0.04 & 0 & 0 & 0 & 0 \\ 0.05 & 0.05 & 0.05 & 0 & 0.05 & 0.04 & 0 & 0 & 0 & 0.06 \\ 0.08 & 0.08 & 0.05 & 0.05 & 0 & 0.04 & 0.25 & 0.07 & 0.07 & 0 \\ 0.03 & 0.07 & 0.04 & 0.04 & 0.04 & 0 & 0.04 & 0.04 & 0.04 & 0 \\ 0.08 & 0.07 & 0 & 0 & 0.25 & 0.04 & 0 & 0 & 0 & 0.33 \\ 0.07 & 0.07 & 0 & 0 & 0.07 & 0.04 & 0 & 0 & 1 & 0.18 \\ 0.07 & 0.16 & 0 & 0 & 0.07 & 0.04 & 0 & 1 & 0 & 0.18 \\ 0 & 0 & 0 & 0.06 & 0 & 0 & 0.33 & 0.18 & 0.18 & 0 \end{bmatrix}$$

We now apply the new clustering procedure, by finding the cohesiveness matrix. The generalized cohesiveness matrix M for the directed fuzzy graph $\vec{G}$ is,

$$M = \begin{bmatrix} 0 & 0.13 & 0.05 & 0.05 & 0.08 & 0.03 & 0.08 & 0.07 & 0.07 & 0.08 \\ 0.13 & 0 & 0.27 & 0.05 & 0.08 & 0.07 & 0.07 & 0.07 & 0.16 & 0.07 \\ 0.05 & 0.27 & 0 & 0.05 & 0.05 & 0.04 & 0.07 & 0.07 & 0.16 & 0.06 \\ 0.05 & 0.05 & 0.05 & 0 & 0.05 & 0.04 & 0.06 & 0.06 & 0.06 & 0.06 \\ 0.08 & 0.08 & 0.05 & 0.05 & 0 & 0.04 & 0.25 & 0.07 & 0.07 & 0.25 \\ 0.03 & 0.07 & 0.04 & 0.04 & 0.04 & 0 & 0.04 & 0.04 & 0.04 & 004 \\ 0.08 & 0.07 & 0.07 & 0.06 & 0.25 & 0.04 & 0 & 0.07 & 0.18 & 0.33 \\ 0.07 & 0.07 & 0.07 & 0.06 & 0.07 & 0.04 & 0.07 & 0 & 1 & 0.18 \\ 0.07 & 0.16 & 0.16 & 0.06 & 0.07 & 0.04 & 0.18 & 1 & 0 & 0.18 \\ 0.08 & 0.07 & 0.05 & 0.06 & 0.25 & 0.04 & 0.33 & 0.18 & 0.18 & 0 \end{bmatrix}$$

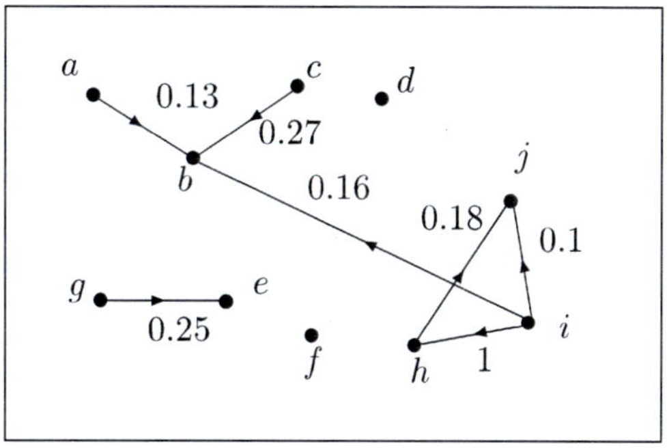

Fig. 9.12 Cluster induced subgraph for the level 0.13

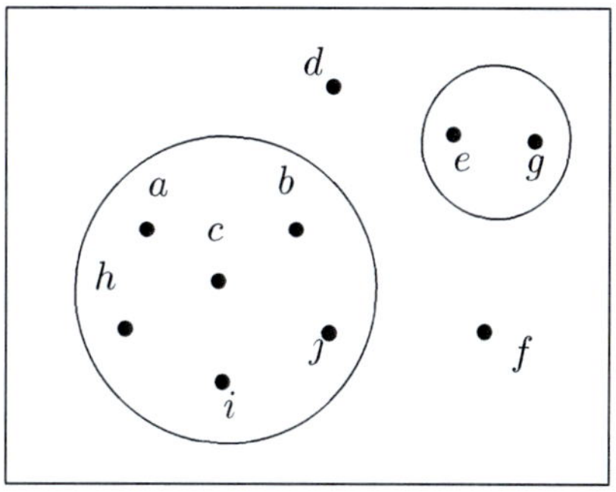

Fig. 9.13 Clusters for the fuzzy subgraph in Fig. 9.12

Using the algorithm from Ref. [17], we have $K'_f(\overrightarrow{G}) = 0.03$. Using the generalized $t-$connectivity procedure, we can have the fuzzy clusters as given bellow.

Level	Fuzzy Clusters
$(1, \infty)$	$\{a\}, \{b\}, \{c\}, \{d\}, \{e\}, \{f\}, \{g\}, \{h\}, \{i\}, \{j\}$
$(0.33, 1]$	$\{d\}, \{f\}, \{a\}, \{b, c\}, \{e\}, \{g\}, \{j\}, \{h, i, \}$
$(0.25, 0.33]$	$\{d\}, \{f\}, \{a\}, \{b, c\}, \{e\}, \{g, j\}, \{h, i, \}$
$(0.18, 0.25]$	$\{d\}, \{f\}, \{a\}, \{b, c\}, \{e, g, j\}, \{h, i\}$
$(0.16, 0.18]$	$\{d\}, \{f\}, \{a\}, \{b, c\}, \{e, g, j\}, \{h, i\}$
$(0.13, 0.16]$	$\{d\}, \{f\}, \{a\}, \{b, c\}, \{e, g, h, i, j\}$
$(0.08, 0.13]$	$\{d\}, \{f\}, \{e, g\}, \{a, b, c, h, i, j\}$
$(0.07, 0.08]$	$\{d\}, \{f\}, \{e, g\}, \{a, b, c, h, i, j\}$
$(0.06, 0.07]$	$\{d\}, \{a, b, c, e, f, g, h, i, j\}$
$(0.05, 0.06]$	$\{a, b, c, d, e, f, g, h, i, j\}$
$(0.04, 0.05]$	$\{a, b, c, d, e, f, g, h, i, j\}$
$(0.03, 0.04]$	$\{a, b, c, d, e, f, g, h, i, j\}$
$(0, 0.03]$	$\{a, b, c, d, e, f, g, h, i, j\}$

The induced subgraph of G with generalized edge connectivity $\kappa'_f(G) = 0.13$ is given in Fig. 9.12.

Clusters of the above subgraph are given in Fig. 9.13.

The induced subgraph H of the directed graph G having $\kappa'_f(H) = 0.18$ is given in Fig. 9.14.

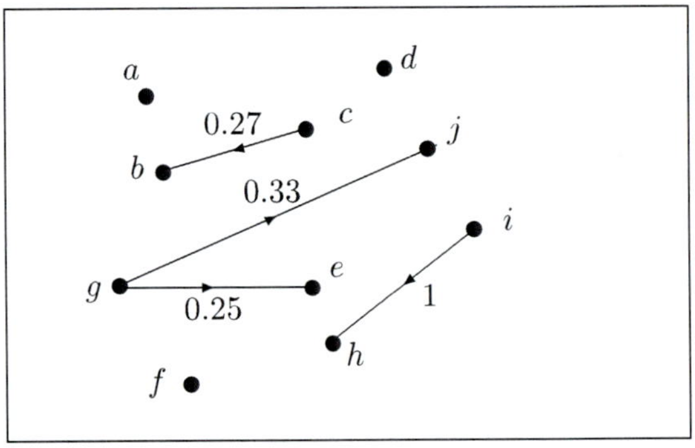

Fig. 9.14 Cluster induced subgraph for the level 0.18

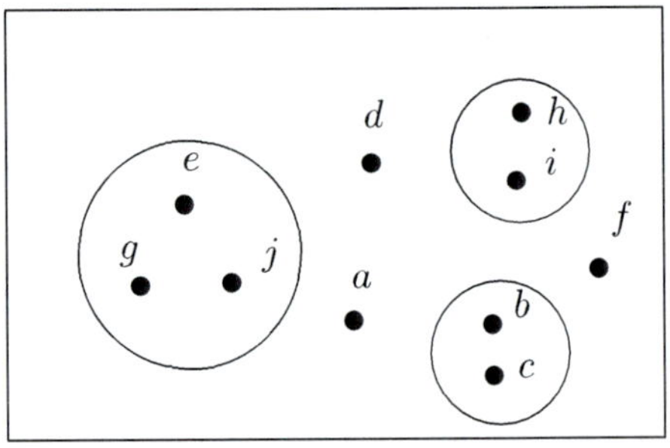

Fig. 9.15 Clusters for the subgraph in Fig. 9.14

The diagram representing clusters of the above subgraph is given in Fig. 9.15.

We can see that there is a moderate flow of humans between most of the locations, except South America, eastern Europe and central Asia. There is a reasonably high flow from the cluster formed by South Asia, Middle East, Sub-saharan Africa to North and Central America and Central Europe to West and Southern Europe. It is possible to assign numerical parameters to the clusters at any level. For example, the average cluster density ρ defined by $\frac{n_1 * k_1 + n_2 * k_2 + \cdots + n_l * k_l}{n_1 + n_2 + \cdots + n_l}$, where n_i denote the number of vertices in ith connected component in the cluster induced subgraph and k_i denote the generalized edge connectivity in the ith connected component, gives value 0.16 for the level $t = 0.13$ and 0.47 for the level $t = 0.18$. More realistic parameters can be defined using cluster size, average distance, and so forth.

Bibliography

1. 2017 Global Estimate of Modern Slavery and Child Labor, Alliance 8.7, Regional brief for Europe and Central Asia
2. Abiyev, R.H.: Fuzzy wavelet neural network based on fuzzy clustering techniques for time series prediction. Neural Comput. Appl. **20**, 249–259 (2011)
3. Ali, S., Mathew, S., Mordeson, J.N., Rashmonolu, H.: Vertex connectivity of fuzzy graphs with applications to human trafficking. New Math. Nat. Comput. **14**(3), 457–485 (2018)
4. Adamo, J.H.: Fuzzy decision tress. Fuzzy Sets Syst. **4**, 207–219 (1980)
5. Ahmed, B., Gani, A.N.: Perfect fuzzy graphs. Bull. Pure Appl. Sci. **28**, 83–90 (2009)
6. Akram, M.: Bipolar fuzzy graphs. Inf. Sci. **181**, 5548–5564 (2011)
7. Akram, M.: Bipolar fuzzy graphs with applications. Knowl.-Based Syst. **39**, 1–8 (2013)
8. Akram, M.: Cofuzzy graphs. J. Fuzzy Math **19**, 1–12 (2011)
9. Alaoui, A.: On fuzzification of some concepts of graphs. Fuzzy Sets Syst. **101**, 363–389 (1999)
10. Complete fuzzy graphs: AL-Hawary, T. Int. J. Math. Combanitorics **4**, 26–34 (2011)
11. Anjali, N., Mathew, S.: Critical blocks in fuzzy graphs. J. Fuzzy Math. **23**(4), 907–916 (2015)
12. Anjali, N., Mathew, S.: On blocks and stars in fuzzy graphs. J. Intell. Fuzzy Syst. **28**, 1659–1665 (2015)
13. Anjali, N., Mathew, S.: Transitive blocks and their applications in fuzzy interconnection networks. Fuzzy Sets Syst. **352**, 142–160 (2018)
14. Anthony, M.C.: A Hidden scourge, international monetary fund. Finance Dev. **55**(3), 19–21 (2018)
15. Atannasov, K.T.: Intuitionistic Fuzzy Sets: Theory and Applications, Studies in Fuzziness and Soft Computing, vol. 35. Springer, Physica-Verlag (1999)
16. Bandler, W., Kohout, L.: Fuzzy power sets and implication. Fuzzy Sets Syst. **4**, 13–30 (1980)
17. Banerjee, S.: An optimal algorithm to find degrees of connectedness in an undirected edge weighted graph. Pattern Recogn. Lett. **12**, 421–424 (1991)
18. Batista, M.: Estimation of the final size of the coroavirus epidemic by the SIR model. Research-Gate 1–14 (2020)
19. Bhattacharya, P.: Some remarks on on fuzzy graphs. Pattern Recogn. Lett. **6**, 297–302 (1987)
20. Bhattacharya, P., Suraweera, F.: An algorithm to compute the supremum of max-min powers and a property of fuzzy graphs. Pattern Recogn. Lett. **12**, 413–420 (1991)
21. Bhutani, K.R.: On automorphisms of fuzzy graphs. Pattern Recogn. Lett. **9**, 159–162 (1989)

J. N. Mordeson and S. Mathew, *Mathematics of Uncertainty for Coping with World Challenges*, Studies in Systems, Decision and Control 353,
https://doi.org/10.1007/978-3-030-68684-0

22. Bhutani, K.R., Battou, A.: On M-strong fuzzy graphs. Inf. Sci. **155**, 103–109 (2003)
23. Bhutani, K.R., Rosenfeld, A.: Fuzzy end vertices in fuzzy graphs. Inf. Sci. **152**, 323–326 (2003)
24. Bhutani, K.R., Rosenfeld, A.: Geodesics in fuzzy graphs. Electronic Notes Discrete Math. **15**, 51–54 (2003)
25. Strong arcs in fuzzy graphs: Bhutani, K.R and Rosenfeld, A. Inf. Sci. **152**, 319–322 (2003)
26. Bhutani, K.R., Mordeson, J.N., Rosenfeld, A.: On degrees of end vertices in fuzzy graphs. Iran. J. Fuzzy Graphs **1**(1), 57–64 (2004)
27. Biglarbegiav, A., Sadeghian, W., Melek, M.: On the accessability/controllability of fuzzy control systems. Inf. Sci. **202**, 58–72 (2012)
28. Binu, M., Mathew, S., Mordeson, J.: Connectivity index of a fuzzy graph and its applications to human trafficking. Fuzzy Sets Syst. **360**, 117–136 (2019)
29. Binu, M., Mathew, S., Mordeson, J.: Wiener index of a fuzzy graph and application to illegal immigration networks. Fuzzy Sets Syst. **384**, 132–147 (2020)
30. Binu, M., Mathew, S., Mordeson, J.: Cyclic connectivity index of fuzzy graphs. IEEE Trans, Fuzzy Syst (2020). (In press)
31. Boulmakoul, A.: Fuzzy graph modelling for HazMat telegeomonitoring. Eur. J. Oper. Res. **175**, 1514–1525 (2006)
32. Boyd, D., Brickell, K., Brown, D., Ives, C., Natarajan, N., Parsons, L., Beaujet, C.: Modern Slavery, Environmental Destruction and Climate Change: Fisheries, Fields, Forests and Factories, A research report from the University of Nottingham's Rights Lab, Rotal Holloway University of London, and the Independent Anti-Slavery Commissioner
33. Chanas, S., Kolodziejczynk, W.: Maximum flow in a network with fuzzy arc capacities. Fuzzy Sets Syst. **8**, 165–173 (1982)
34. Chartrand, G., Zhang, P.: Distance in graphs - taking the long view. AKCE Int. J. Graphs Comb. **1**(1), 1–13 (2004)
35. Cheng, C.T., Ou, C.P., Chau, K.W.: Combining a fuzzy optimal model with a generic algorithm on solving multiobjective rainfall-runoff model calibration. J. Hydrol. **268**, 72–86 (2002)
36. Climate Change Evidence, Impacts, Choices answers to common questions about the science of climate change, National Research Council of the National Academies. www.national-academies.org
37. Dahir, A.L.: Africa is again the world's epicenter of modern slavery, Quartz Africa. Accessed 23 July 2018
38. Darbian, E., Borzooei, R.A.: Results on vague graphs with applications in human trafficking. New Math. Nat. Comput. **14**, 37–52 (2018)
39. Delgado, M., Verdegay, J.L., Vila, M.A.: On fuzzy tree definition. Eur. J. Oper. Res. **22**, 243–249 (1985)
40. Dempster, A.P.: Upper and lower probability inferences based on a sample from a finite univariate population. Biometrika **54**(3, 4), 515–528 (1967)
41. Dempster, A.P.: Upper and lower probabilities induced by multivalued mappings. Ann. Math. Stat. **38**(2), 325–339 (1967)
42. Deo, N.: Graph Theory with Applications to Engineering and Computer Science. Prentice Hall, Englewood Cliffs (1990)
43. Deschrijver, G., Cornelis, C.: Representability in interval-valued fuzzy set theory. Int. J. Uncert. Fuzziness Knowl.-Based Syst. **15**, 345–361 (2007)
44. Diamond, P.: A fuzzy max-flow min-cut theorem. Fuzzy Sets Syst. **119**, 139–148 (2001)
45. Dirac, G.A.: Some theorems on abstract graphs. Proceed. Lond. Math. Soc. **3**(1), 69–81 (1952)
46. Dubois, D., Prade, H.: Fuzzy Sets and Systems. Academic, New York (1980)
47. Dudley, S.: Human Trafficking in the Middle East and North Africa, Topical Research Digest: Human Rights and Human Trafficking, pp. 74–87
48. Eckstein, D., Hutfils, M.L., Winges, M.: Global Climate Risk Index 2019, Who Suffers Most From Extreme Weather Events? Weather-related Loss Events in 2017 and 1998 to 2017, Germanwatch

49. Erdos, P., Gallai, A.: On maximal paths and circuits in graphs. Acta Math. Sci. Hung **10**, 337–356 (1959)
50. Fang, J.F.: The bipancycle-connectivity of hypercube. Inf. Sci. **178**, 4679–4687 (2008)
51. Frank, A.: Think: Opinion, Analysis, Essays (2020)
52. Gacto, M.J., Alcala, R., Herrera, F.: Interpretability of Linguistic fuzzy rule-based systems: an overview of interpretability measures. Inf. Sci. **181**(20), 4340–4360 (2011)
53. Gagne, D.: Organized Crime Profits from Modern Slavery in Latin America. Insight Crime. Accessed 3 June 2016
54. Gani, A.L., Radha, K.: Isomorphism on fuzzy graph. Int. J. Comput. Math. Sci. **2**, 190–196 (2008)
55. Gerrard, M.B.: Climate Change and Human Trafficking After the Paris Agreement, pp. 345–368
56. Global Food Security Index 2019, Economist Intelligence Unit, Supported by Corteva Agriscience https://foodsecurityindex.eiu.com/
57. Global Hunger Index. https://www.globalhungerindex.org/results.html (2019)
58. Global Report on Trafficking in Persons, United Nations Office on Drugs and Crime: Vienna, Austria (2014)
59. Global Slavery Index. http://www.globalslaveryindex.org/findings/ (2016)
60. Global Slavery Index 2018 Walk Free Foundation. www.globalslaveryindex.org
61. Global Slavery Index, Europe and Central Asia (2019)
62. Goldman, A.J., Witzgall, C.J.: A localization theorem for optimal facility location. Trans. Sci. **4**, 406–409 (1970)
63. Guiasu, S.: Reaching a verdict by weighting evidence. In: Wang, P. (ed.) Advances in Fuzzy Theory and Technology, vol. II, pp. 167–180. Durham, North Carolina (1994)
64. Harary, F.: Graph Theory (Third printing). Addison Wesley, Boston (1972)
65. Homenda, W., Pedrycz, W.: Balanced fuzzy gates. RSCTC, 107–116 (2006)
66. Huber, K.P., Berthold, M.R.: Application of fuzzy graphs for metamodeling. In: Proceedings, 1998. IEEE World Congress on Computational Intelligence. Anchorage, USA. Accessed 4–9 May 1998 (current version 6 Aug 2002)
67. Hubbs, C.: Social distancing to slow the coroavirus: Modeling the flattening of the COVID19 peaks, Towards Data Science, https://towardsdatascience.com/social-distancing-to-slow-thecoronavirus (2020)
68. ICAT - The Inter-Agency Coordination Group Against Trafficking in Persons, The Role of the Sustainable Development Goals (SDGs) in Combating Trafficking in Persons (2018)
69. Jicy, N., Mathew, S.: Connectivity analysis of cyclically balanced fuzzy graphs. Fuzzy Inf. Eng. **7**, 245–255 (2015)
70. Kaufman, A.: Introduction a la Theorie ds sons-ensembles flous, vol. 1, pp. 41–189. Masson, Paris (1973)
71. Kauffmann, A.: Introduction to the Theory of Fuzzy Sets, vol. 1. Academic, Orlando (1973)
72. Kiss, A.: An application of fuzzy graphs in database theory. PU. M. A. Ser. A **1**, 337–342 (1990)
73. Kiss, A.: λ-decomposition of fuzzy relational databases. Ann. Univ. Sci. Budapest Sect. Comp. **12**, 133–148 (1991)
74. Klein, C.M.: Fuzzy shortest paths. Fuzzy Sets Syst. **39**, 27–41 (1991)
75. Koczy, L.T.: Fuzzy graphs in the evaluation and optimization of networks. Fuzzy Sets Syst. **46**, 307–319 (1992)
76. Kolinjivadi, V.: Opinion/Coronavirus Pandemic, The coronavirus outbreak is part of the climate change crisis (2020)
77. Ksaifi, L.: Modern Slavery in the Middle East, Australian Institute of International Affairs. Accessed 23 June 2016
78. Lam, Y.F., Li, V.: Reliability modeling and analysis of communication networks with dependent failures. IEEE Trans. Commun. **34**, 82–84 (1986)
79. Lee, K.M.: Bipolar-valued fuzzy sets and their basic operations. In: Proceedings of the International Conference, Bangkok, Thailand (2000), pp. 307–317

80. Lin, J.Y., Cheng, C.T., Chau, K.W.: Using support vector machines for long-term discharge prediction. Hydrolog. Sci. J. **51**, 599–612 (2006)
81. Lin, C.K., Huang, H.M., Tan, J.J., Hsu, L.H.: On spanning connected graphs. Discrete Math. **308**(7), 1330–1333 (2008)
82. Lyth, A.: DW News on Facebook, Why Southeast Asia struggles to tackle modern slavery, 9/4 (2015)
83. Malik, D.S., Acharjee, S., Mathew, S., Mordeson, J.: Soft fuzzy correlation coefficient: application to illegal immigration. J. Fuzzy Math. **27**(1), 63–76 (2019)
84. Malik, D.S., Mordeson, J.N.: Fuzzy Discrete Structures, Studies in Fuzziness and Soft Computing, vol. 58. Physica-Verlag, Heidelberg (2000)
85. Malik, D.S., Mathew, S., Mordeson, J.: N, Fuzzy incidence graphs: Applications to Human Trafficking. Inf. Sci. **447**, 244–255 (2018)
86. Mandal, S., Bhatnagar, T., Arinaminpathy, N., Agarwal, A., Chowdhury, A., Murhekar, M., Gangakhedkar, R.R., Sarkar, S.: Prundent public health intervention strategies to control the coronavirus disease 2019 transmission in India: A mathematical model-based approach. Indian J. Med. Res. E-pub ahead of print https://doi.org/10.4103/ijmr.IJMR_504_20
87. Marczewski, E.: Sur deux Proprieties des Classes d'ensembles. Fund. Math. **33**, 303–307 (1945)
88. Mathew, J.K., Mathew, S.: On strong extremal problems in fuzzy graphs. J. Intell. Fuzzy Syst. **30**, 2497–2503 (2016)
89. Mathew, J.K., Mathew, S.: Some special sequences in fuzzy graphs. Fuzzy Inf. Eng. **8**, 31–40 (2016)
90. Mathew, S., Mordeson, J.N., Malik, D.S.: Fuzzy Graph Theory. Springer, New York (2018)
91. Mathew, S., Mordeson, J.N.: Directed fuzzy networks as a model to illicit flows and Max-Flow Min-Cut theorem (SI on Human Trafficking). New Math. Nat. Comput. **13**(3), 219–229 (2017)
92. Mathew, S., Mordeson, J.N.: Fuzzy incidence blocks and their applications to illegal migration problems (SI on Human Trafficking). New Math. Nat. Comput. **13**(3), 245–260 (2017)
93. Mathew, S., Yang, H.L., Mathew, J.K.: Saturation in fuzzy graphs. New Math. Nat. Comput. **14**(1), 113–128 (2018)
94. Mathew, S., Sunitha, M.S.: A characterization of blocks in fuzzy graphs. J. Fuzzy Math. **18**(4), 999–1006 (2010)
95. Mathew, S., Sunitha, M.S.: Bonds in graphs and fuzzy graphs. Adv. Fuzzy Sets Syst. **6**(2), 101–119 (2010)
96. Mathew, S., Sunitha, M.S.: Cycle connectivity in fuzzy graphs. J. Intell. Fuzzy Syst. **24**, 549–554 (2013)
97. Mathew, S., Sunitha, M.S.: Menger's theorem for fuzzy graphs. Inf. Sci. **222**, 717–726 (2013)
98. Mathew, S., Sunitha, M.S.: Node connectivity and arc connectivity in fuzzy graphs. Inf. Sci. **180**, 519–531 (2010)
99. Mathew, S., Sunitha, M.S.: Strongest strong cycles and θ-fuzzy graphs. IEEE Trans. Fuzzy Syst. **21**(6), 1096–1104 (2013)
100. Mathew, S., Sunitha, M.S.: Types of arcs in a fuzzy graph. Inf. Sci. **179**(11), 1760–1768 (2009)
101. Mathew, S., Sunitha, M.S., Anjali, N.: Some Connectivity concepts in bipolar fuzzy graphs. Ann. Pure Appl. Math. **7**, 98–108 (2014)
102. Matula, D.W.: Cluster analysis via graph theoretic techniques. In: Proceedings of the Louisiana Conference on Combinatorics, Graph Theory, and Computing, pp. 199–212 (1970)
103. Mendel, J.M.: Uncertainty Rule-Based Fuzzy Logic Systems: Introduction and New Directions. Prentice-Hall, Upper Saddle River (2001)
104. Mendal, J.M., Gang, X.: Fast computation on centroids for constant-width interval-valued fuzzy sets, pp. 621–626. Fuzzy Information Processing Society, NAFIPS (2006)
105. Miyamoto, S.: Fuzzy Sets in Information Retieval and Cluster Analysis, Theory and Decision Library, Series D: System Theory, Knowledge Engineering and Problem Solving. Kluwer Academic Publishers (1990)

106. Mordeson, J.N.: Fuzzy line graphs. Pattern Recogn. Lett. **14**, 381–384 (1993)
107. Mordeson, J.N., Mathew, S.: Human trafficking: Source, transit, destination, designations (SI on Human Trafficking). New Math. Nat. Comput. **13**(3), 209–218 (2017)
108. Mordeson, J.N., Mathew, S.: Local look at human trafficking (SI on Human Trafficking). New Math. Nat. Comput. **13**(3), 327–340 (2017)
109. Mordeson, J.N., Mathew, S.: Non Deterministic flow in fuzzy networks and its application in identification of human trafficking chains (SI on Human Trafficking). New Math. Nat. Comput. **13**(3), 231–243 (2017)
110. Mordeson, J.N., Mathew, S.: Advanced Topics in Fuzzy Graph Theory. Springer, Berlin (2019)
111. Mordeson, J.N., Mathew, S.: Sustainable goals in combating human trafficking: Analysis by mathematics of uncertainty. J. Algebr. Hyperstructures Log Algebras **1**(1), 49–59 (2020)
112. Mordeson, J.N., Mathew, S.: Fuzzy Logic Applied to Sustainable Development Goals and Human Trafficking. Symmetry **12**, 87 (2020). https://doi.org/10.3390/sym12010087
113. Mordeson, J.N., Mathew, S.: Sustainable Development Goals: Analysis by Mathematics of Uncertainty. Springer, Berlin (2020). (In Press)
114. Mordeson, J.N., Mathew, S., Aacharjee, S.: Aggregation operators: applications to human trafficking and slavery. New Math. Nat. Comput. **14**(3), 403–421 (2018)
115. Mordeson, J.N., Mathew, S., Binu, M.: Dialectic synthesis: application to human trafficking. New Math. Nat. Comput. **15**(3), 395–410 (2019)
116. Mordeson, J.N., Mathew, S., Ali, S.: Fuzzy path fans applied to health consequences of trafficking victims. Adv. Fuzzy Sets Syst. **24**(1), 17–28 (2019)
117. Mordeson, J.N., Mathew, S., Borzooei, R.A.: Vulnerability and government response to human trafficking: vague Fuzzy incidence graphs. New Math. Nat. Comput. **14**(2), 203–219 (2018)
118. Mordeson, J.N., Mathew, S., Malik, D.S.: Fuzzy Graph Theory with Applications to Human Trafficking. Springer, Berlin (2018)
119. Mordeson, J.N., Mathew, S., Mallenby, M., Aacharjee, S.: Human trafficking: policy intervention (SI on Human Trafficking). New Math. Nat. Comput. **13**(3), 341–358 (2017)
120. Mordeson, J.N., McCoy, A.S., Mathew, S., Binu, M.: Fuzzy indices with applications to human trafficking. New Math. Nat. Comput. **15**(3), 411–425 (2019)
121. Mordeson, J.N., Mordeson, J.H., Mathew, S.: Directed graphs applied to human trafficking. New Math. Nat. Comput. **14**(3), 445–455 (2018)
122. Mordeson, J.N., Nair, P.S.: Arc Disjoint Fuzzy Graphs. IEEE 0-7803-5211-4/99 (1999)
123. Mordeson, J.N., Nair, P.S.: Cycles and cocycles of fuzzy graphs. Inf. Sci. **90**, 39–49 (1996)
124. Mordeson, J.N., Nair, P.S.: Fuzzy Graphs and Fuzzy Hypergraphs. Physica - Verlag, Heidelberg (2000)
125. Mordeson, J.N., Nair, P.S.: Fuzzy Mathematics. Physica-Verlag, Heidelberg (2001)
126. Mordeson, J.N., Peng, C.S.: Operations on fuzzy graphs. Inform. Sci. **79**, 159–170 (1994)
127. Mordeson, J.N., Schwab-McCoy, A., Mathew, S., Binu, M.: Fuzzy indices with applications to human trafficking. New Math. Nat. Comput. **15**, 411–425 (2019)
128. Mordeson, J.N., Yao, Y.Y.: Fuzzy cycles and fuzzy trees. J. Fuzzy Math. **10**, 189–202 (2002)
129. Mordeson, J.N., Wething, H.C., Clark, T.C.: A fuzzy mathematical model of nuclear stability. New Math. Nat. Comput. **6**, 119–140 (2010)
130. Nagoor Gani, A., Latha, S.R.: A new algorithm to find fuzzy Hamilton cycle in a fuzzy network using adjacency matrix and minimum vertex degree. Springer Plus **5**(1), 18–54 (2016)
131. New Insights in Climate Science: A 2017–2019 Summary, Future Earth, Produced by Future Earth and the Earth League for the United Nations Climate Action Summit (2019)
132. OECD Responsible Business Conduct: OECD Guidelines for Multicultural Enterprise, vol. 12., Paris (2019)
133. Osborn, D., Cutter, A., Ullah, F.: Universal Sustainable Development Goals, pp. 1–25. Report of a Study by Stakeholder Forum, Stakeholder Forum, May, Understanding the Transformational Challenge for Developed Countries (2015)
134. Paun, A.: Fragile Planet. Scoring climate risks around the world, Climate Change Global HSBC, Global Research March (2018)
135. Pedrycz, W.: Fuzzy Sets Engineering. CRC Press, Boca Raton (1995)

136. Pedrycz, W.: Human centricity in computing with fuzzy sets: an interpretability quest for higher granular constructs. J. Ambient Intell. Hum. Comput. **1**, 65–74 (2010)
137. Pedrycz, W., Bargiela, A.: Fuzzy clustering with semantically distinct families of variables, descriptive and predictive aspects. Pattern Recogn. Lett. **31**, 1952–1958 (2010)
138. Perchant, A., Bloch, I.: Fuzzy morphisms between graphs. Fuzzy Sets Syst. **128**, 149–168 (2002)
139. Rickard, C.: Human Trafficking Report Shows sub-Saharan Africa a Global Player, African Lii. Accessed 18 July 2019
140. Riff, M., Zúniga, M., Montero, E.: A graph-based immune-inspired constraint satisfaction search. Neural Comput. Appl. **19**, 1133–1142 (2010)
141. Roberts, F.: Discrete Mathematical Models. Prentice Hall, Englewood Cliffs (1976)
142. Robinson, J.R., Mordeson, J.N., Mathew, S.: Foster care deficiencies and human trafficking. New Math. Nat. Comput. **15**(2), 215–230 (2019)
143. Rosenfeld, A.: Fuzzy graphs. In: Zadeh, L.A., Fu, K.S., Shimura, M. (eds.) Fuzzy Sets and Their Applications, pp. 77–95. Academic, New York (1975)
144. Routledge Taylor and Francis Group: Real-world impact. CRC Press, Clairvate Analytics (2019)
145. Roy, M.K., Biswas, R.: I-V fuzzy relations and Sanchez's approach to medical diagnosis. Fuzzy Sets Syst. **47**, 35–38 (1992)
146. Saaty, T.L.: A scaling method for priorities in hierarchical structures. J. Math. Psychol. **15**, 234–281 (1977)
147. Saaty, T.L.: The Analytic Hierachy Process. McGraw Hill, New York (1980)
148. Sachs, Schmidt-Traub, G., Kroll, C., Durand-Delacre, D., Teksoz, K.: SDG Index and Dashboards Report 2017: Global Responsibilities International, Spillovers in Achieving the Goals, Sustainable Development Solutions Network, A Global Initiative for the United Nations, New York Bertelsmann Stiftung and Sustainable Development Solutions Network (SNDSN), , pp. 1–479 (2017)
149. Sachs, J., Schmidt, G., Kroll, C., Lafortune, G., Fuller, G.: Sustainable Developments Report 2019. Bertelsmann Stiftung and Sustainable Development Solutions Network (SDSN), New York (2019)
150. Sameena, K., Sunitha, M.S.: Strong arcs and maximum spanning trees in fuzzy graphs. Int. J. Math. Sci. 17–20 (2006)
151. Arya, S., Mordeson, J.N., Mathew, S.: Generalized fuzzy graph connectivity parameters with application to human trafficking. Mathematics **8**, 424 (2020). https://doi.org/10.3390/math8030424
152. SDG Alliance 8.7, Joining forces globally to end child labour, forced labour, modern slavery and human trafficking, International Labour Organization
153. Shafer, G.: A Mathematical Theory of Evidence. Princeton University Press, Princeton (1976)
154. Shannon, A., Atanassov, K.T.: A first step to a theory of the intuitionistic fuzzy graph. In: Lakov, D. (ed.) Proceeding of FUBEST, Sofia, pp. 59-61. Accessed 28–30 Sept 1994
155. Shannon, A., Atanassov, K.T.: Intuitionistic fuzzy graphs from α-, β-, and (α, β)-levels. Notes Intuit. Fuzzy Sets **1**, 32–35 (1995)
156. Smith, D., Moore, L.: The SIR model for spread of disease - The differential equation model. Math. Assoc. Amer. 1–5 (2020)
157. Sunitha, M.S., Vijayakumar, A.: A characterization of fuzzy trees. Inf. Sci. **113**, 293–300 (1999)
158. Sunitha, M.S., Vijayakumar, A.: Blocks in fuzzy graphs. J. Fuzzy Math. **13**(1), 13–23 (2005)
159. Sunitha, M.S., Vijayakumar, A.: Complement of a fuzzy graph. Indian J. Pure Appl. Math. **33**, 1451–1464 (2002)
160. Sunitha, M.S., Mathew, S.: Fuzzy graph theory, a survey. Ann. Pure Appl. Math. **4**(1), 92–110 (2013)
161. Tamura, S., Higuchi, S., Tanaka, K.: Pattern classification based on fuzzy relations. IEEE Trans. SMC-1, 61–66 (1971)

162. Tong, Z., Zheng, D.: An algorithm for finding the connectedness matrix of a fuzzy graph. Congr. Numer. **120**, 189–192 (1996)
163. Trafficking in Persons: Global Patterns, The United Nations Office for Drugs and Crime, Trafficking in Persons Citation Index
164. Trafficking in Persons Report - Cuba, United States Department of State, 28 (2018)
165. United Catholic Sisters Against Human Trafficking, The surprising link between climate change and human trafficking, The Revelator (2019)
166. Wang, X., De Baets, B., Kerre, E.: A comparative study of similarity measures. Fuzzy Sets Syst. **73**, 259–268 (1995)
167. Wu, S.Y.: The composition of fuzzy graphs. J. Res. Educ. Sci. **31**, 603–629 (1986)
168. Wu, C.L., Chau, K.W., Li, Y.S.: Predicting monthly stream flowusing data driven models coupled with data pre-processing techniques. Water Resour. Res. **45**, W08432 (2009). https://doi.org/10.1029/2007WR006737
169. Xie, J.X., Cheng, C.T., Chau, K.W., Pei, Y.Z.: A hypbrid adaptive time-delayed neural network model for multi-step-ahead prediction of sunspot activity. Int. J. Env. Pollut. **28**, 364–381 (2006)
170. Xu, J.: The use of fuzzy graphs in chemical structure research. In: Rouvry, D.H. (ed.) Fuzzy Logic in Chemistry, pp. 249–282. Academic, New York (1997)
171. Yeh, R.T., Bang, S.Y.: Fuzzy graphs, fuzzy relations, and their applications to cluster analysis. In: Zadeh, L.A., Fu, K.S., Shimura, M. (eds.) Fuzzy Sets Appl., pp. 125–149. Academic, New York (1975)
172. Yen, J.: Generalizing the Dempster-Shafer theory to fuzzy sets. In: Wang, Z., Klir, G.J. (eds.) Fuzzy Measure Theory (Chap. 7), pp. 257–283. Plenum Press, New York (1992)
173. Yener, B., Gunduz, C., Gultekin, S.H.: The cell graphs of cancer. Bioinformatics **20**(1), 145–151 (2004)
174. Yin, Y., Zhan, J., Corsini, P.: Fuzzy roughness of n-ary hypergroups based on a complete residuated lattice. Neural Comput. Appl. **20**, 41–57 (2011)
175. Zadeh, L.A.: From precise to granular probabilities. Fuzzy Sets Syst. **154**, 370–374 (2005)
176. Zadeh, L.A.: Fuzzy sets. Inform. Control **8**, 338–353 (1965)
177. Zadeh, L.A.: Is there a need for fuzzy logic? Inf. Sci. **178**(13), 2751–2779 (2008)
178. Zadeh, L.A.: Similarity relations and fuzzy orderings. Inf. Sci. **3**, 177–200 (1971)
179. Zadeh, L.A.: Toward a generalized theory of uncertainty (GTU) an outline. Inf. Sci. **172**, 1–40 (2005)
180. Zadeh, L.A.: The concept of a linguistic and application to approximate reasoning. Inf. Sci. **8**, 199–249 (1975)
181. Zhang, J., Chau, K.W.: Multilayer ensemble pruning via novel multi-sub-swarm particle swarm optimization. J. Univ. Comput. Sci. **15**(4), 840–858 (2009)
182. Zhang, J., Yang, X.: Some properties of fuzzy reasoning in propositional fuzzy logic systems. Inf. Sci. **180**, 4661–4671 (2010)

Index

J. N. Mordeson and S. Mathew, *Mathematics of Uncertainty for Coping with World Challenges*, Studies in Systems, Decision and Control 353,
https://doi.org/10.1007/978-3-030-68684-0